INTERNATIONAL DEVELOPMENT IN FOCUS

Air Pollution and Climate Change

From Co-Benefits to Coherent Policies

Grzegorz Peszko, Markus Amann, Yewande Awe, Gary Kleiman, and Tamer Samah Rabie

 **WORLD BANK GROUP**

Contents

Boxes

Figures

Tables

Acknowledgments

Financial support from the Pollution Management and Environmental Health Programme (PMEH) is gratefully acknowledged. This report was written and edited by Grzegorz Peszko with substantive contributions from Yewande Awe (chapter 1), Tamer Samah Rabie (chapters 1 and 2), and Gary Kleiman (chapter 3). Martin Heger, Gordon Hughes, and Rodrigo Pizarro also contributed substantively, as acknowledged in the text. Markus Amann was a mentor, contributor, critical reviewer, and great adviser throughout the process of developing this report. Benoit Blarel ignited the idea of writing this report and guided its infancy. Karin Kemper, Christian Albert Peter, and Ernesto Sánchez-Triana provided guidance and encouragement in its maturity. The following peer reviewers helped correct mistakes and improve the narrative: Paulina Estela Schulz Antipa, Simon Black, Stephane Hallegate, Martin Heger, Dafei Huang, Gary Kleiman, Muthukumara Mani, Marcelo Mena, Ernesto Sánchez-Triana, and Xiaodong Wang. Much guidance and inspiration that influenced this volume was generously offered by Sameer Akbar, Anna Dworakowska, Andrzej Gula, David Hanrahan, Solvita Klapare, Zbigniew Klimont, Kseniya Lvovsky, Urvashi Narajan, Jostein Nygard, Klas Sander, Karin Shepardson, Qing Wang, Marek Zaborowski, Vasil Zlatev, and many others. Professional editing was carried out by Stan Wanat, and the World Bank Publishing Program produced the book.

Abbreviations

AAP	Ambient air pollution
AQM	air quality management
BAT	Best Available Techniques
BC	Black carbon
CAFE	Corporate Average Fuel Economy
CAIR	Clean Air Interstate Rule
CCAC	Climate and Clean Air Coalition
CFC	Chlorofluorocarbons
CHP	Combined heat and power
EEA	European Environmental Agency
ERF	Effective radiative forcing
ESP	electrostatic precipitator
ETD	Energy Taxation Directive
ETS	Emissions Trading System
EU	European Union
FGD	Flue-gas desulfurization
GAHP	Global Alliance on Health and Pollution
GAINS	Gas-air pollution interactions and synergies
GDP	Gross domestic product
GHG	greenhouse gas
HAP	Household air pollution
HDV	Heavy-duty vehicles
IAQCC	Integrated air quality and climate change
ICCI	International Cryosphere Climate Initiative
IEA	International Energy Agency
IHME	Institute for Health Metrics and Evaluation
IIASA	International Institute for Applied Systems Analysis
IMF	International Monetary Fund
IPCC	Intergovernmental Panel on Climate Change
LNG	Liquefied natural gas
LPG	Liquefied petroleum gas
NDC	Nationally determined contributions
NO_x	nitrogen oxides
OC	Organic carbon

OECD	Organisation for Economic Co-operation and Development
PM	Particulate matter
PMEH	Pollution Management and Environmental Health
PMF	Positive Matrix Factorization
SCC	Social cost of carbon
SCR	Selective catalytic reduction
SLCP	Short-lived climate pollutants
UNEP	United Nations Environment Programme
VOC	Volatile organic compounds
WHO	World Health Organization
WMO	World Meteorological Organization

Executive Summary

INTRODUCTION

This report analyzes the experiences of countries and communities trying to address both air pollution and climate change. It proposes a coherent policy approach to harnessing synergies and managing tensions between these two essential challenges to sustainable development. The coherent, or integrated, approach to air quality management and climate change policies discussed in this report is a dynamic policy process that puts people's health first while paving the way for long-term decarbonization.

The conceptual framework of this report is based on the concept of *policy coherence,* which is defined as an approach to integrating the dimensions of sustainable development throughout domestic and international policy making (OECD 2019a). *Policy integration* is further defined by the Organisation for Economic Co-operation and Development (OECD) as "a process by which institutions align their mandates, policies and sectoral objectives ... , and make their policy decisions taking into account the interactions (synergies and trade-offs) among economic, social and environmental areas" (OECD 2019a).

This is not an advocacy report. It assumes that the reader understands why it is important to solve these two existential challenges to humanity and is not still wondering whether they are real. This book offers evidence-based practical guidance on how to align air quality management and climate-mitigation policies to get things done effectively and efficiently on the ground.

This report serves as a reality check to the belief that one environmental problem will be solved as a co-benefit of the other. Despite several synergies between policies and measures to address both problems, there are at least a few major tensions. Climate-mitigation policies, applied in isolation, can temporarily deteriorate air quality. On the other hand, air pollution policies applied without corresponding climate policies can temporarily warm the planet and also lock in large carbon emissions in combustion plants equipped with expensive end-of-pipe air pollution controls.

Decarbonization is a long-term structural transformation that will take decades, especially in developing and emerging economies where most people exposed to poor air quality live. In the meantime, 7 million lives can be saved every year by reducing population exposure to air pollution, including from sources that are still burning fossil fuels. This report aids in the design of policy processes that will help prioritize pollutants, emission sources, and measures that quickly save peoples' lives and navigate the journey to a low-carbon future. It encourages policy makers to recognize synergies and trade-offs transparently and manage them with coherent and integrated policies.

POLLUTANTS

Economic activities often involve emitting several air pollutants and greenhouse gases (GHGs) simultaneously. Some air pollutants warm the climate, while others cool it (table ES.1). A number of gases, called short-lived climate pollutants—such as black carbon, tropospheric ozone, carbon monoxide, and volatile organic compounds—cause air pollution and warm the planet at the same time. Other air pollutants—such as sulfur dioxide (SO_2), nitrogen oxides (NO_x), ammonia (NH_3), organic carbon, and secondary inorganic aerosols—cool the earth with a stronger countervailing force. These pollutants have cancelled roughly one-third of the total global warming resulting from GHG emissions since the 1800s. Likewise, the GHG with the largest cumulative warming potential, carbon dioxide (CO_2), has no effect on air pollution, and the second-most-important GHG—methane—has an indirect effect on air pollution as a precursor of ground-level ozone (see table ES.1).

TABLE ES.1 Impact of pollutants on local human health (through air pollution) and climate change

POLLUTANT	LOCAL HEALTH IMPACT	CLIMATE IMPACT	CO-BENEFITS OR TRADE-OFFS BETWEEN AIR POLLUTION AND CLIMATE CHANGE
1. Black carbon—component of $PM_{2.5}$	Harmful	Warming	Synergy between air pollution and climate (short-lived climate pollutants)
2. Ground-level ozone (O_3)	Harmful	Warming	
3. Methane (CH_4)	Harmful indirectly	Warming	
4. Carbon monoxide (CO)	Harmful	Warming	
5. Volatile organic compounds (VOCs)	Harmful	Warming	
6. Organic carbon (OC)	Harmful	Cooling	Trade-offs between air pollution and climate mitigation
7. Sulfur dioxide (SO_2)	Harmful	Cooling	
8. Nitrogen oxides (NO_x)	Harmful	Cooling	
9. Ammonia (NH_3)	Harmful	Cooling	
10. Secondary inorganic aerosols	Harmful	Cooling	
11. Heavy metals, benzo[a]pyrene, dioxins, and so on	Harmful	Neutral	
12. Carbon dioxide (CO_2)	Neutral	Warming	Long-term climate forcers, neutral for air quality
13. Chlorofluorocarbons (CFCs)	Neutral	Warming	
14. Hydrofluorocarbons (HFCs)	Neutral	Warming	
15. Nitrous oxide (N_2O)	Neutral	Warming	

Source: World Bank.
Note: $PM_{2.5}$ = particulate matter two-and-one-half microns or less in width.

EMISSION SOURCES

Global and local pollutants are often co-emitted from the same sources, but the same sources are rarely equal priorities for air pollution and climate change.

Urban transport, agriculture, and waste management account for the greatest overlap of priority sources of air pollution and GHGs, but rarely are they the main culprits of either problem.

Among stationary combustion installations, small sources burning solid fuels, including biomass, coal, and waste, are the main sources of population exposure to air pollution. These sources include household heating and cooking, artisanal boilers and stoves, and open and uncontrolled burning of municipal or agricultural waste. Among them, biomass energy sources are part of a solution to climate change as renewable energy. Consequently, those who want to quickly save peoples' lives from air pollution would focus on small, dispersed emission sources.

In contrast, quick and cost-effective reduction of emissions of the most potent GHGs can be achieved by shutting down large thermal industrial and power plants, decarbonizing long-distance transport, and switching to a plant-based diet. Large installations also co-emit large amounts of SO_2 and NO_x, which are climate coolants. The impact of these large combustion sources on local air quality can be significant in specific airsheds but often is secondary because emissions come from high stacks and locations where pollutants are dispersed farther away from locally situated exposed populations. Again, addressing sources of short-lived climate pollutants offers synergies between air pollution and climate mitigation, especially medium-term and local temperature peaking, but they may or may not be the critical and most cost-effective sources with which to start abatement in a particular airshed.

Source location, local weather patterns, and the individual characteristics of emission sources have critical impacts on air pollution but have irrelevant impacts on climate change (except black carbon, whose climate forcing is localized).

EMISSIONS-ABATEMENT MEASURES

Next, synergies and trade-offs between technical and behavioral measures to address air quality and climate change should be addressed. Most emissions-control measures affect both, but they do not always improve both environmental issues. Emissions-abatement measures with strong climate-air pollution synergies include shutting down the least-efficient obsolete plants, improving energy efficiency, and undertaking better operation and maintenance of polluting installations.

Once the potential of such measures is exhausted, the trade-offs between air quality and climate mitigation become more common. Further major strides toward clean air can be achieved quickly and relatively cheaply by retrofitting existing plants and vehicles with end-of-pipe pollution-control equipment or by switching household heating and cooking from biomass to natural gas or electricity from power plants fired by fossil fuels. Such measures have played a crucial role in reducing air pollution in China, India, and OECD countries with no impact, or a small negative impact, on climate (figure ES.1). Installations such

FIGURE ES.1

Cost-effectiveness of measures to reduce concentration of air pollutants (PM$_{2.5}$ or O$_3$)

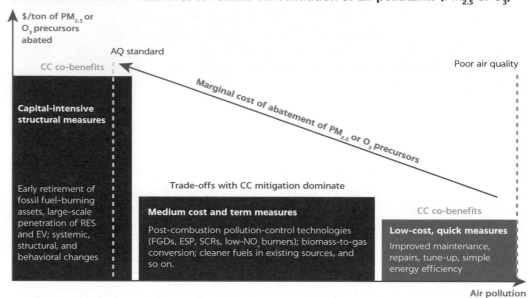

Source: World Bank.
Note: AQ = air quality; CC = climate change; ESP =electrostatic precipitator; EV = electric vehicle; FGDs = flue-gas desulfurization; NO$_x$ = nitrogen oxides; O$_3$ = ozone; PM$_{2.5}$ = particulate matter two-and-one-half microns or less in width; RES = renewable energy sources; SCR = selective catalytic reduction.

as particulate filters, flue-gas desulfurization units, and catalytic reduction systems for NO$_x$ need electricity and materials to operate. Therefore, these measures increase the on-site use of fossil fuels and reduce the useful energy produced from a unit of fuel. This leads to higher CO$_2$ emissions per unit of output of plants or per kilometer driven by vehicles, to say nothing of sometimes significantly higher capital and operating costs of the entire plant. Therefore, when energy or carbon prices go up and air pollution policies are weak, plant operators sometimes reduce the operating hours of air pollution controls or uninstall them altogether, leading to dire health consequences. This report reviews examples and conditions under which this happens and identifies policy approaches to prevent it.

Nowhere does the trade-off have larger health implications than for household cooking and heating in developing countries. Indoor and ambient air pollution caused by burning solid fuels, mainly biomass, in small stoves at family homes kills between 3 and 4 million people every year (Health Effects Institute 2020). Children and women are disproportionately exposed because they spend more time at and near home. Often, the cheapest and most effective way to quickly improve the health of families is to switch from biomass to bottled liquefied petroleum gas, natural gas, or electricity and district heating fired by fossil fuels. Development finance institutions are reluctant to support such measures because biomass is a renewable energy and gas is a fossil fuel. On the other hand, higher energy and carbon prices also induce lower-income households to switch from "cleaner" fossil fuels back to polluting biomass.

The local and temporary trade-offs between air pollution reduction and climate change mitigation are as common as win-win abatement measures

FIGURE ES.2

Synergies and potential tensions between key mitigation measures for air pollution and climate change

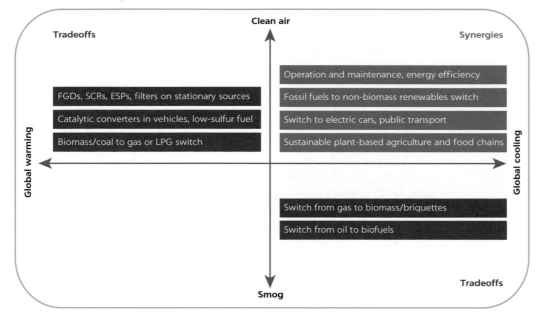

Source: World Bank.
Note: ESPs = electrostatic precipitators; FGDs = flue-gas desulfurization units; LPG = liquefied petroleum gas; SCR = selective catalytic reduction.

available to economic agents, as illustrated in figure ES.2. Their presence is a fact that must be managed and not an excuse to delay action on either air pollution or climate change.

High-income economies can afford leapfrogging directly to the accelerated large-scale phaseout of fossil fuels. However, in developing countries, local communities often struggle with more limited resources and capabilities with which to simultaneously address multiple environmental problems. Carbon-free technologies may yet be unaffordable or unreliable and require extensive infrastructure, such as power grids, electric vehicle charging, or electricity storage to replace fossil fuels. Consequently, if such local communities must choose, they tend to prioritize measures that quickly prevent premature deaths over measures that lead to a lasting transition to a low-carbon future. The net climate-warming effect of these health-driven air pollution measures is usually small, temporary, and well justified by the lives saved, especially in countries that contribute very little to global warming.

In the opposite direction, there is a risk that the notorious reliance on end-of-pipe equipment to control air pollution can lead to the excessive accumulation of capital in carbon-intensive installations and thereby extend their lifespan. The good news for the climate is that investments in state-of-the-art air pollution controls also make fuel-combustion plants, equipment, and vehicles more expensive to build and operate, undermining their cost competitiveness compared with nonfossil alternatives. However, this can happen only in energy markets that are competitive and face both high carbon pricing and stringent air pollution regulations. These conditions are still rare outside the OECD countries.

INTEGRATED AIR QUALITY AND CLIMATE CHANGE POLICY PROCESS

Climate policies can induce the "first mile" of air quality improvements by encouraging simple energy efficiency improvements and can induce the "last mile" by stranding any remaining combustion sources in the future. However, in several current circumstances, climate policies can temporarily increase the emission of air pollutants when air pollution policies are weak or poorly targeted, for example, when high fuel prices (1) discourage the operation of technologies to control air pollution and (2) make cleaner fossil fuels unaffordable and encourage fuel switching to biomass, waste, or low-quality coal distributed through informal markets. In the opposite direction, stringent air pollution policies in the presence of low fuel prices may temporarily increase GHG emissions by locking in large amounts of capital in carbon-intensive assets.

Synergies and trade-offs need to be recognized and managed by coherent and integrated public policies. This report recommends that targeted climate and air quality regulations always be implemented jointly and calibrated to harness win-win opportunities when relevant and mitigate the risk of aggravating one environmental problem while solving another. There is no evidence that climate policies implemented so far have materially improved air quality when and where needed; however, there is strong evidence that targeted air pollution policies have had a significant impact on the health of exposed populations. Therefore, an integrated air quality and climate change (IAQCC) approach dynamically focuses on the near-term health impacts of air pollution in the most polluted airsheds while paving the way for the long-term phaseout of fossil fuels. A *coherent and integrated* policy framework enables plant operators and households to adjust their decisions, jointly considering air pollution and low-carbon goals. An IAQCC always begins at the airshed level and typically consists of the five steps illustrated in figure ES.3.

1. In the first step of the IAQCC policy process, a ground-level system for monitoring air quality is established to determine where indoor air pollution or ambient air pollution, such as tropospheric ozone and particulate matter two-and-one-half microns or less in width ($PM_{2.5}$), reaches concentrations that pose a risk to health.

2. The second step examines the exposure of the population, assets, and ecosystems to poor air quality in pollution hot spots and estimates the impact it has on health and other damage. Monetary valuation of these impacts can follow, if appropriate. This assessment underpins a choice of targets for air quality improvement. Agencies responsible for attaining these targets are identified (or established, if needed), roles and responsibilities are assigned, and a broader institutional and governance framework is created to enable implementation of the subsequent steps.

3. The third step centers on identification of the key emission sources responsible for population exposure to harmful pollutants, especially to $PM_{2.5}$ and ozone (O_3) in the targeted airsheds. In this step, climate mitigation goals can be integrated into air quality management processes.

    –   Source-apportionment studies use laboratory tests of samples of $PM_{2.5}$ suspended in the ambient air to identify the type of emission source they

FIGURE ES.3

The five steps of the integrated air quality and climate change policy process

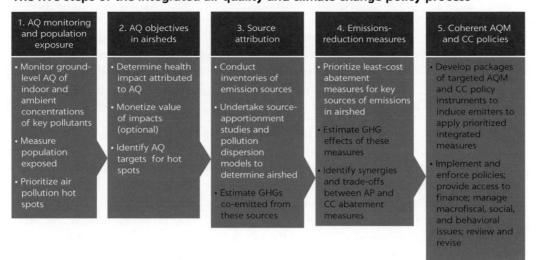

1. AQ monitoring and population exposure	2. AQ objectives in airsheds	3. Source attribution	4. Emissions-reduction measures	5. Coherent AQM and CC policies
• Monitor ground-level AQ of indoor and ambient concentrations of key pollutants • Measure population exposed • Prioritize air pollution hot spots	• Determine health impact attributed to AQ • Monetize value of impacts (optional) • Identify AQ targets for hot spots	• Conduct inventories of emission sources • Undertake source-apportionment studies and pollution dispersion models to determine airshed • Estimate GHGs co-emitted from these sources	• Prioritize least-cost abatement measures for key sources of emissions in airshed • Estimate GHG effects of these measures • Identify synergies and trade-offs between AP and CC abatement measures	• Develop packages of targeted AQM and CC policy instruments to induce emitters to apply prioritized integrated measures • Implement and enforce policies; provide access to finance; manage macrofiscal, social, and behavioral issues; review and revise

Source: World Bank.
Note: White text denotes actions driven primarily by health considerations. Black text refers to actions that gradually introduce an integrated approach to mitigating air pollution and climate change. AP = air pollution; AQ = air quality; AQM = air quality management; CC = climate change; GHG = greenhouse gas.

came from, such as vehicles, households, industries, waste burning, agriculture, or nutrients—but not necessarily the location of the sources. Such studies include the identification of sources of direct emissions of $PM_{2.5}$—and importantly, emissions of their precursors, such as SO_2, NO_x, volatile organic compounds, and ammonia. Source-apportionment studies can also be generated by models, such as the greenhouse gas-air pollution interactions and synergies (GAINS) model, but the model's results need to be validated by laboratory studies.

- Inventories of relevant emission sources need to be established to map source location, capacity, load profile, type, amount of fuel use, and emissions of air pollutants (if monitored), as well as key source characteristics such as age, stack height, combustion technology, equipment, timing of emissions, cropping patterns, agricultural technologies, and so on.

- Models of airshed pollution dispersion trace the exact locations from which the pollutants came and their atmospheric formations so that policy interventions can be targeted at the key sources of emissions that contribute to population exposure in the most polluted location.

4. The fourth step involves an assessment of costs of available technical, behavioral, and structural abatement measures, as well as the potential of these measures to reduce the exposure of the population in hot spots. The impact of the most cost-effective measures to abate air pollution on climate forcing can be estimated at this stage. Synergies and trade-offs between priority measures to mitigate air pollution and climate change can be identified. Measures with high climate co-benefits can be prioritized if they do not significantly compromise air quality. Additional climate-mitigation efforts may be identified if the priority air pollution measures temporarily increase GHG emissions (for example, installation of end-of-pipe control measures or switch from biomass to gas) or if these climate-mitigation efforts reduce

emissions of climate coolants (such as SO_2 and NO_x). The assessment of choices should consider the incremental health effects, premature deaths, incremental costs, capacity, and financing constraints, as well as the social and distributional impacts of prioritizing air pollution measures with climate co-benefits. Decisions on capital-intensive end-of-pipe air pollution measures, should be supported by an assessment of the economic and social risks of possible future premature retirement of fossil fuel assets and associated contingent fiscal liabilities.

5. The fifth step is to design, implement, and enforce the integrated package of regulations to incentivize firms and households to implement the abatement measures prioritized earlier. In this fifth step, the mix of policy instruments needs to encourage economic agents to change investment and behavioral decisions considering both the short-term cost of air pollution and the long-term cost of climate change. Integration of policies means a creative design of comprehensive mixes of direct regulations (such as emissions-performance standards, requirements for the use of the best available techniques, or zoning requirements) along with economic and fiscal instruments that would give firms and households adequate flexibility to achieve air quality objectives at the least cost and encourage innovation and discovery of new, creative abatement measures. If air pollution policies induce abatement measures that increase GHG emissions, then additional climate policy efforts need to be identified. Likewise, strengthening air pollution instruments will be necessary if more ambitious climate policy instruments (for example, carbon prices) risk increasing air pollution and adverse health impacts. Calibrating the mix, and the level of ambition, of air pollution and climate policy instruments through a dynamic process of reviews and adjustments must take into account the local conditions and political economy dynamics discussed in this report.

ALIGNING INCENTIVES WITH FINANCE

This report shows that it is essential for firms and households to face coherent policy incentives to address both air pollution and climate change. These instruments need to be tuned to local and global impacts to optimize choices, such as whether to reduce pollution from fossil fuel installations or leapfrog to new technologies free of fossil fuels. These choices can differ by airshed and the health hazards to the population exposed to air pollution.

Affordability, social and distributional impact, and access to finance are also important factors shaping choices, especially in low- and middle-income countries. International and bilateral development institutions face the challenge of rethinking their financing policies and their assistance to developing countries to proactively manage the synergies and trade-offs between the risks of air pollution and climate change. Always prioritizing assistance to win-win measures, especially in regard to domestic heating and cooking, may lead to preventable diseases and premature deaths from air pollution without materially helping the climate agenda.

REFERENCES

Health Effects Institute. 2020. "State of Global Air 2020: A Special Report on Global Exposure to Air Pollution and Its Health Impacts." Health Effects Institute, Boston, MA. https://www.stateofglobalair.org/.

OECD (Organisation for Economic Co-operation and Development). 2019a. "Recommendation of the Council on OECD Legal Instruments Policy Coherence for Sustainable Development." OECD/LEGAL/0381. OECD, Paris. https://www.oecd.org/gov/pcsd/recommendation-on-policy-coherence-for-sustainable-development-eng.pdf.

OECD (Organisation for Economic Co-operation and Development). 2019b. *Taxing Energy Use 2019: Using Taxes for Climate Action*. Paris: OECD Publishing. https://doi.org/10.1787/058ca239-en.

World Bank. 2017. *Toward a Clean World for All: An IEG Evaluation of the World Bank Group's Support to Pollution Management*. Sector and Thematic Evaluation. Washington, DC: World Bank Group. https://hubs.worldbank.org/docs/imagebank/pages/docprofile.aspx?nodeid=28830447.

1 Do Air Pollution and Climate Matter for Health?

HEALTH IMPACTS OF AIR POLLUTION

The Global Burden of Disease *Lancet* study ranks ambient air pollution (AAP) as the fourth-largest level-2 health risk factor for global deaths in 2019 (Global Health Metrics 2020). The World Health Organization (WHO) estimates almost 4.2 million people died prematurely from AAP and almost 3.8 million people died from household air pollution (HAP) in 2016.[1] More than 90 percent of the world's population lives in places where air quality exceeds WHO guideline limits.[2] Pollution hot spots are located mainly in the South Asia and Middle East and North Africa regions, where particulate matter ($PM_{2.5}$) concentrations are about eight to nine times higher than in North America and the exposed population is much larger. Air-pollution exposure is also high in East Asia and Pacific, dominated by China, and in Eastern Europe.

Air pollution exerts the greatest burden of disease of all environmental factors (Landrigan et al. 2018), with significant impacts on health from both short-term (Mills et al. 2015) and long-term exposure (Hoek et al. 2013). Exposure to poor quality air is associated with increased mortality and increased risks of developing cardiovascular and respiratory diseases and certain cancers.

Breaking mortality down by disease cause shows the following: For AAP, ischemic heart disease is the largest cause of death. For HAP, acute lower respiratory disease and ischemic heart disease account for substantially equal numbers of deaths (table 1.1).

Air pollution is also associated with many other detrimental, but less researched, health impacts and conditions, such as low birthweight (Ezziane 2013), preterm delivery, diabetes (Bowe et al. 2018), mental health conditions (Shin, Park, and Choi 2018), and neurological impairment (Xu, Ha, and Basnet 2016), including dementia in later life (Carey et al. 2018). As the evidence base for these and other conditions becomes ever stronger, it should be possible to include an even larger true health burden from air pollution in global estimates.

Air pollution hurts the vulnerable and poor the most. Babies and those under five years old suffer most from HAP. Among adults, men have greater mortality from AAP whereas HAP affects both genders similarly. This finding reflects the

TABLE 1.1 Estimated deaths attributable by disease to ambient and household air pollution, 2016

Thousands

DISEASE	AMBIENT AIR POLLUTION	HOUSEHOLD AIR POLLUTION
Acute lower respiratory disease	754	1,006
Chronic obstructive pulmonary disease	765	763
Ischemic heart disease	1,589	1,031
Lung cancer	262	285
Stroke	827	686
All causes	**4,197**	**3,771**

Source: WHO, "Burden of Disease from AAP for 2016" (http://www.who.int/airpollution/data/AAP_BoD _results_May2018_final.pdf).

relatively greater time that men spend outside the home. The poorest households are typically more exposed and more vulnerable, that is, least able to escape from polluted zones and protect themselves with air purification or other expensive treatment.

HEALTH IMPACTS OF CLIMATE CHANGE

Unlike air pollutants, which are directly harmful to people when inhaled in concentrations exceeding WHO guideline values, climate pollutants are harmless to people in concentrations that are harmful to the climate. Yet, climate pollutants cause global warming, which is associated with more frequent and intense occurrence of extreme weather events that are harmful to humans and increase probability of some diseases, such as malaria. Therefore, the link between emissions of greenhouse gases and health is indirect and delayed. The best documented health risks of climate change include[3] the following:

- Disaster-related health impacts are likely to increase with the intensification of cyclones and floods.
- Heat stress will worsen as high temperatures become more common and water scarcity increases.
- Malnutrition, especially in children, is projected to become more prevalent, with an increase in droughts and where livelihoods are threatened by coastal erosion or warming seas.
- Vector-borne and water-borne diseases will expand in range as conditions favor mosquitoes, flies, and other sources of pathogens.

These impacts will result in greater risks of injury, disease, and death as well as lost work capacity and labor productivity (Ebi, Campbell-Lendrum, and Wyns 2018). Air pollution and climate change can mutually aggravate each of their health hazards when combined. Large populations in developing countries are more susceptible to the effects of extreme weather events such as heat waves when they are also exposed to poor air quality (Schnell 2017). A strong line of evidence also suggests that most changes to regional meteorology attributed to climate change are conducive to buildup of local pollutants (Fiore et al. 2012; Jacob and Winner 2009; Kirtman et al. 2013; Silva et al. 2013).

ECONOMIC VALUE OF BENEFITS OF IMPROVED AIR QUALITY: PUTTING A PRICE TAG ON HEALTH

The health effects of ambient $PM_{2.5}$ exposure can be monetized to provide an estimate of its social cost. Valuation of mortality follows the welfare approach, or value of statistical life, in World Bank and IHME (2016). Morbidity, measured as days of illness, is valued at wage rates, but morbidity constitutes a minor part of total health costs.

Local air pollutants account for an annual global welfare loss estimated at $4.6 trillion (Landrigan et al. 2018) to $5.11 trillion (World Bank and IHME 2016), which is equivalent to more than 6 percent of global gross domestic product (GDP). The welfare cost is highest in South Asia (equivalent to 7.3 percent of GDP) and in upper-middle-income countries (equivalent to 6 percent of GDP), and lowest in the Americas and low-income countries. It is worth stressing that comparison to GDP is done only for illustrative purposes, because, in reality, diseases boost GDP by increasing demand for monetary transactions involving pharmaceuticals, and health care services. Premature deaths due to air pollution reduce GDP if the pollution affects people during the years that they are of working age. For example, in 2013 the global economy lost about $225 billion of labor income.

Some cities and countries use economic valuation of the health benefits of air quality improvement to strengthen the case for action. This approach is common in the European Union and the United States. The World Bank has supported several cost-of-environmental-degradation studies to calculate the costs of premature mortality and morbidity associated with air pollution. In Tehran, for example, the World Bank–led research (Heger and Sarraf 2018) indicates that AAP is associated with the premature deaths of about 4,000 people per year. It was further estimated that reducing $PM_{2.5}$ levels to those in London would prevent about 1,300 premature deaths per year in Tehran, and reducing $PM_{2.5}$ levels to those in New York City (15 micrograms per cubic meter) would prevent about 2,000 premature deaths per year in Tehran. These avoided premature deaths translate into monetary savings for the city of roughly US$1 billion a year if London air pollution levels were achieved, and more than US$1.5 billion a year if New York City levels of air pollution were reached (figure 1.1).

Shindell (2015) assesses co-benefits and trade-offs between local and global pollution in the US power and transportation sectors, introducing the Social Cost of Atmospheric Release measure. This metric expands the damage function typically used for the social cost of carbon (SCC) calculations (expanding the damage function typically used for SCC calculation and focusing on well-mixed greenhouse gases), and also includes the value of damage caused by local air pollutants and short-lived climate pollutants. He finds that a broader view of these sectors' environmental damage for 2015 US electricity generation would amount to about US$0.14–US$0.34 per kilowatt-hour for coal, about US$0.04–US$0.18 for natural gas, US$3.80 per gallon of gasoline, and US$4.80 per gallon of diesel.

FIGURE 1.1

Annual economic benefits of reducing PM$_{2.5}$ concentrations in Tehran

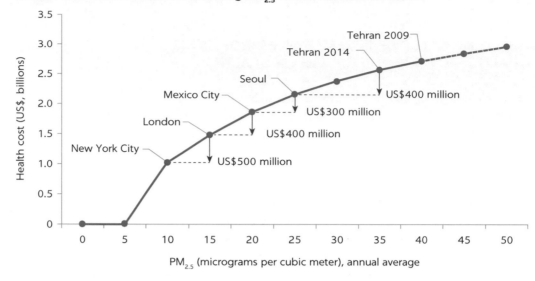

Source: Heger and Sarraf 2018.

Note: PM$_{2.5}$ = particulate matter two-and-one-half microns or less in width.

NOTES

1. Air pollutants, depending on how they are formed, fall into two broad categories: primary pollutants and secondary pollutants. Both contain chemical compounds in solid, liquid, or vapor phases. Key pollutants that are harmful to public health and the environment include PM (both primary and secondary), tropospheric ozone (secondary), nitrogen oxides, sulfur oxides, carbon monoxide, lead, and other heavy metals, and toxic air pollutants. Air pollution can be of natural origins or anthropogenic origins, primarily combustion processes.
2. "WHO air pollution" (http://www.who.int/airpollution/en/ accessed November 10, 2019.)
3. "Climate Change and Health in IDA-Supported Countries" (http://pubdocs.worldbank .org/en/621451516135118033/FINAL-IDA-Hotspot-Note-9-Jan-2018.pdf).

REFERENCES

Bowe, B., Y. Xie, T. Li, Y. Yan, H. Xian, and Z. Al-Aly. 2018. "The 2016 Global and National Burden of Diabetes Mellitus Attributable to PM$_{2.5}$ Air Pollution." *Lancet Planetary Health* 2 (7): 301–12. doi:10.1016/S2542-5196(18)30140-2.

Carey, Ian M., H. Ross Anderson, Richard W. Atkinson, Sean D. Beevers, Derek G. Cook, David P. Strachan, David Dajnak, John Gulliver, and Frank J. Kelly. 2018. "Are Noise and Air Pollution Related to the Incidence of Dementia? A Cohort Study in London, England." *BMJ Open* 8: e022404. doi:10.1136/bmjopen-2018-022404.

Ebi, Kristie, Diarmid Campbell-Lendrum, and Arthur Wyns. 2018. "The 1.5 Health Report: Synthesis on Health and Climate Science in the IPCC SR1.5." World Health Organization, Geneva. http://www.who.int/globalchange/181008_the_1_5_healthreport.pdf?ua=1.

Ezziane, Z. 2013. "The Impact of Air Pollution on Low Birth Weight and Infant Mortality." *Review of Environmental Health* 28 (2–3): 107–15. doi:10.1515/reveh-2013-0007.

Fiore, Arlene M., Vaishali Naik, Dominick V. Spracklen, Allison Steiner, Nadine Unger, Michael Prather, Dan Bergmann, et al. 2012. "Global Air Quality and Climate." *Chemical Society Reviews* 41: 6663–83. doi:10.1039/c2cs35095e.

Global Health Metrics. 2020. "Global Burden of 87 Risk Factors in 204 Countries and Territories, 1990–2019: A Systematic Analysis for the Global Burden of Disease Study 2019." *The Lancet* 396 (10258): 1223–49.

Heger, Martin, and Maria Sarraf. 2018. "Air Pollution in Tehran: Health Costs, Sources, and Policies." Environment and Natural Resources Global Practice Discussion Paper 6, World Bank, Washington, DC. https://openknowledge.worldbank.org/handle/10986/29909.

Hoek, G., R. M. Krishnan, R. Beelen, A. Peters, B. Ostro, B. Brunekreef, and J. D. Kaufman. 2013. "Long-Term Air Pollution Exposure and Cardio-Respiratory Mortality: A Review." *Environmental Health* 12 (1): 43. doi:10.1186/1476-069X-12-43.

Jacob, D. J., and D. A. Winner. 2009. "Effect of Climate Change on Air Quality." *Atmospheric Environment* 43 (1): 51–63. doi:10.1016/j.atmosenv.2008.09.051.

Kirtman, Ben, Scott B. Power, Akintayo John Adedoyin, George J. Boer, Roxana Bojariu, Ines Camilloni, Francisco Doblas-Reyes, et al. 2013. "Near-Term Climate Change: Projections and Predictability." In *Climate Change 2013: The Physical Science Basis, Contribution of Working Group I to the Fifth Assessment Report of the Intergovernmental Panel on Climate Change*, 953–1028. Cambridge, UK, and New York, NY: Cambridge University Press.

Landrigan, P. J., R. Fuller, N. J. R. Acosta, O. Adeyi, R. Arnold, N. N. Basu, A. B. Baldé, et al. 2018. "The Lancet Commission on Pollution and Health." *Lancet* 391 (10119): 462–512.

Mills I. C., R. W. Atkinson, S. Kang, H. Walton, and H. R. Anderson. 2015. "Quantitative Systematic Review of the Associations between Short-Term Exposure to Nitrogen Dioxide and Mortality and Hospital Admissions." *BMJ Open* 5: e006946. doi:10.1136/bmjopen-2014-006946.

Schnell, Jordan L., and Michael J. Prather. 2017. "Co-Occurrence of Extremes in Surface Ozone, Particulate Matter, and Temperature over Eastern North America." The Proceedings of the National Academy of Sciences (PNAS). https://doi.org/10.1073/pnas.1614453114.

Shin, J., J. Y. Park, and J. Choi. 2018. "Long-Term Exposure to Ambient Air Pollutants and Mental Health Status: A Nationwide Population-Based Cross-Sectional Study." *PLoS One* 13 (4): e0195607. doi:10.1371/journal.pone.0195607.

Shindell, DT. 2015. "The Social Cost of Atmospheric Release." *Climatic Change* 130 (2): 313–326. doi.org/10.1007/s10584-015-1343-0.

Silva, R. A., J. J. West, Y. Zhang, S. C. Anenberg, J.-F. Lamarque, D. T. Shindell, W. J. Collins, et al. 2013. "Global Premature Mortality Due to Anthropogenic Outdoor Air Pollution and the Contribution of Past Climate Change." *Environmental Research Letters* 8 (3): 034005. doi:10.1088/1748-9326/8/3/ 034005. (See https://iopscience.iop.org/article/10.1088/1748 -9326/8/3/031002 for correction.)

World Bank and IHME (Institute for Health Metrics and Evaluation). 2016. *The Cost of Air Pollution: Strengthening the Economic Case for Action*. Washington, DC: World Bank.

Xu, X., S. U. Ha, and R. Basnet. 2016. "A Review of Epidemiological Research on Adverse Neurological Effects of Exposure to Ambient Air Pollution." *Frontiers in Public Health* 4: 157. doi:10.3389/fpubh.2016.00157.

2 Incentives to Reduce Emissions: Economists' Perspective

WHOSE COSTS AND WHOSE BENEFITS?

Environmental policy is justified in economic terms if its total benefits outweigh the total costs. Economic efficiency is maximized when emissions are reduced to the level at which the marginal cost of abatement (the cost of removing the last ton of pollutant) equals the marginal benefit of avoided damages. Because different benefits and costs materialize at different times in the future, the decision-makers should compare their discounted net present values.

Cost-benefit assessments that included monetary valuation of health benefits and avoided premature deaths informed the European Union Clean Air Strategy (Commission of the European Communities 2005). The World Health Organization (WHO) air quality guidelines, which are underpinned by the precautionary principle, were also taken into consideration. In the final political agreement, the European Union (EU) member states agreed on standards for ambient air quality that are less stringent than the WHO guidelines, but linked to the WHO interim targets. The technical feasibility and the costs of attainment of the guideline value recommended by the WHO was deemed temporarily higher than the associated marginal benefits.

Those who bear the costs of pollution reduction are not always those who benefit from it. Improved air quality is enjoyed almost immediately after abatement has occurred and primarily by the region or country that did the abatement. However, air pollutants, including the short-lived climate pollutants, also defy the borders of individual properties, communities, countries, and even continents given that some of them stay in the air for several weeks or months[1] and can travel long distances. This situation prompted countries of the Northern Hemisphere to sign the Convention on Long-Range Transboundary Air Pollution to contain emissions of sulfur dioxide (SO_2), nitrogen oxides (NO_x), volatile organic compounds, and other gases that cause acid rain and eutrophication that harms fragile ecosystems hundreds of kilometers from the sources of pollution. These external effects of pollution also apply to the finest particulate matter ($PM_{2.5}$), the particles so tiny that they can float with the wind for up to 10 days. Figure 2.1 illustrates that in almost all

FIGURE 2.1

Spatial origin of PM$_{2.5}$ in Indian states, population weighted

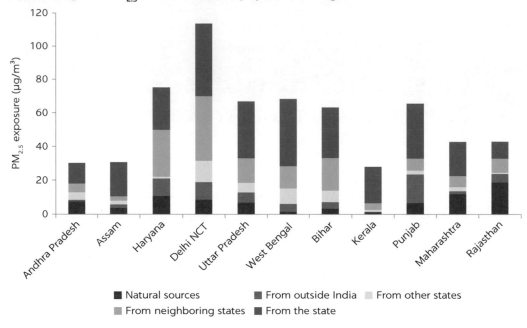

Source: World Bank forthcoming source apportionment study for Indian States conducted by Markus Amann et. al.

Note: NCT = National Capital Territory; PM$_{2.5}$ = particulate matter two-and-one-half microns or less in width; µg/m³ = micrograms per cubic meter.

Indian states, more than half of population exposure to PM$_{2.5}$ originates from other states and even from outside India.

Figure 2.2 shows that European cities have the same problem. Even if the Netherlands completely eliminated local and national emissions of PM$_{2.5}$, the pollution incoming from neighboring countries would still exceed the WHO air quality guidelines. In Poland—a larger country—national PM$_{2.5}$ emissions-reduction efforts should be sufficient to meet WHO air quality guidelines, but individual cities could not do it alone.

Climate change is an extreme example of the geographical and intertemporal divide between polluters and their victims. The benefits of climate change mitigation are long term, less certain, and accrue largely to firms and individuals in other countries. They also transcend generations. Emitted to the atmosphere, greenhouse gases (GHGs) mix evenly around the globe, stay there for hundreds or thousands of years, and cause damage across space and time, including to people who are not yet born or too young to have a say in decision-making today.

Therefore, climate change poses the largest coordination challenge. Almost everybody (though to vastly varying degrees) is both a source and a victim of emissions. Once in the atmosphere, a molecule of GHG warms the climate for everyone. Furthermore, the individual contribution of any single emitter is negligible compared with the stock of pollution in the atmosphere causing global warming. Therefore, today's emitters feel like they are bearing all the current costs of emissions reduction while enjoying only a negligible fraction of associated future benefits. Figure 2.3 illustrates that the private benefit curve for climate change mitigation lies well above the private benefits of improving local

FIGURE 2.2

Individual cities acting alone cannot clean their air

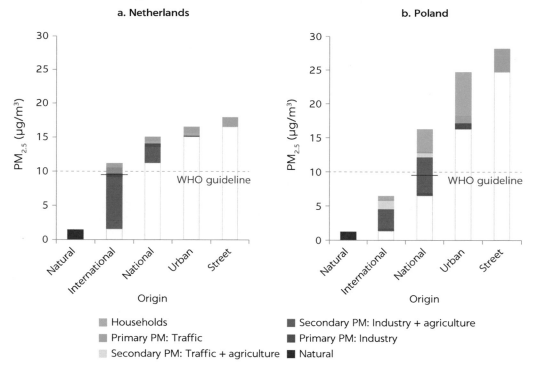

Sources: Kiesewetter and Amann 2014.

Note: PM$_{2.5}$ = particulate matter two-and-one-half microns or less in width; WHO = World Health Organization.

FIGURE 2.3

Air pollution control makes economic sense until the marginal benefit of reducing exposure to pollutants exceeds the marginal cost of doing so

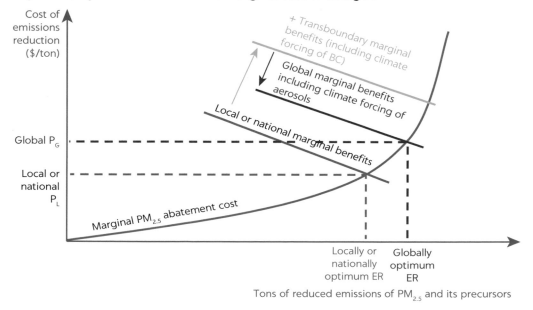

Source: World Bank.

Note: BC = black carbon; ER = reduction of direct PM$_{2.5}$ emissions and its precursors (tons); P$_G$ = global optimum tax (price) on emissions of PM$_{2.5}$ and its precursors; P$_L$ = local or national optimum tax (price) on emissions of PM$_{2.5}$ and its precursors; PM$_{2.5}$ = particulate matter two-and-one-half microns or less in width.

air quality. Polluters have even weaker incentives to be the primary movers on climate action and to be the first to bear the full cost of reducing their GHG emissions, amid concern that other emitters would "free ride" on their effort and enjoy all the benefits of slower warming without contributing to achieving it beyond what is in their private self-interest. Therefore, parties with quantitative mitigation targets under the Kyoto Protocol struggled to enforce them, let alone agree with other countries on allocating shares of emissions reduction efforts to everyone. The Paris Agreement finally gave up on that goal and recognized the primacy of self-interest in climate policies. The incentives to increase the level of action should be created by bottom-up initiatives of clubs of countries bound by similar self-interests.

MICROECONOMICS OF THE COORDINATION CHALLENGE IN PM$_{2.5}$ EMISSIONS CONTROL

It makes economic sense to reduce emissions so long as the marginal benefits of avoided damage remain higher than the marginal costs of mitigation. However, the key question is, whose damages and whose costs? For decision-makers and citizens, the local, private costs of reducing emissions are justified as long as the local marginal benefits of doing so are higher than local marginal costs, that is, until the "Locally optimum ER" in figure 2.3. However, because each particle travels to other towns and countries, causing both health damage and climate warming (as black carbon) and cooling (as organic carbon), the marginal benefits of reducing emissions of one ton of PM$_{2.5}$ are higher to the whole society than the benefits to the local communities that host emitters. The world is interested in deeper emissions reduction, up to the "Globally optimum ER." The marginal cost of achieving this level of emissions reduction is equal to "Global P." A tax on one ton of PM$_{2.5}$ emissions equal to the "Local P$_L$" would be in the self-interest of the local or national community. Higher emissions taxes (such as Global P$_G$) would be justified by joined local and global benefits.

Figure 2.3 illustrates the coordination challenge of solving the air pollution problem. From an emitting city or country perspective, the economically efficient tax on PM$_{2.5}$ emissions is P$_L$ (as suggested by Pigou [1920]) because for the citizens and firms of this city or country, the costs of reducing emissions beyond this level are higher for each ton than the benefits they enjoy. When pollution travels between countries, there is no supranational authority to protect the interests of the distant victims of pollution who are located in other countries or cities. Governments need to engage in multilateral negotiations and international agreements to tackle the problem. For example, the Convention on Long-Range Transboundary Air Pollution was initiated by the Nordic countries, which were the main victims of acid rain and eutrophication caused by pollution originated mainly in the United Kingdom and continental Europe.

The transboundary benefits (the blue schedule in figure 2.3) exceed the local benefits of air quality when health damage, acidification, and eutrophication abroad caused by national emissions of PM$_{2.5}$ and its precursors (such as SO$_2$, NO$_x$, or ammonia) are also accounted for. This curve can also include the global benefits of the reduced near-term climate warming effect of a fraction of PM$_{2.5}$ called orange carbon. The scope of these additional benefits should be reduced (to the red schedule in figure 2.3) by including the climate warming effect of reducing emissions of SO$_2$ and NO$_x$. These air pollutants form atmospheric aerosols that are bad for air quality but disperse sunlight and hence cool the earth (see discussion on gases in chapter 3).

THE HEALTH SECTOR CAN SAVE MILLIONS OF LIVES AND BILLIONS OF DOLLARS BY HELPING ADDRESS THE ROOT CAUSES OF DISEASES RELATED TO AIR POLLUTION

The health care sector bears a large share of the costs of dealing with the negative impacts resulting from air pollution. However, few countries have calculated the health care costs of air pollution. The United Kingdom has estimated its health care costs due to $PM_{2.5}$ and NO_2. These estimates looked at stroke, asthma, lung cancer, chronic obstructive pulmonary disease, diabetes, and low birthweight. In England alone, the related expenditures in the health and social care sectors may have reached £157 million in 2017 and may amount to a further £5.56 billion between 2017 and 2025 (Public Health England 2018). Chile estimated its costs of treating air pollution–related diseases in 2011 at US$9 million per year. The greatest hospital admission costs (US$2.8 million) were found to be due to cardiovascular disease. Emergency admissions from acute bronchitis were estimated to be the greatest overall cost, accounting for more than half of the total figure (US$4.8 million). The rest was the cost of treating asthma, chronic obstructive pulmonary disease, and pneumonia. The total welfare costs of air pollution were found to be about US$670 million per year for the wider economy when lost productivity was taken into account (Chile, Ministerio del Medio Ambiente 2013).

Conservative estimates of the global annual expenditure of the health sector on conditions[2] related to air pollution amount to US$187 billion (Preker et al. 2016). Most of this spending (90 percent) is in high-income countries as a result of the higher absolute spending seen in these wealthier nations (see table 2.1). Although a smaller absolute share of global spending on air pollution is seen in low- and middle-income countries, the share of spending in household income is higher. This potentially avoidable spending is an even more important cost for the already resource-poor health systems in low-income countries, which are least able to afford this expenditure.

The "defensive" expenditures for treating the effects of air pollution can be avoided by improving air quality. Avoided health system costs could be diverted to more productive uses in the economy, creating value to society. In the

TABLE 2.1 **Health-sector expenditures on conditions related to air pollution, 2013**

US$, billions

COUNTRY INCOME LEVEL	AMBIENT AIR POLLUTION	HOUSEHOLD AIR POLLUTION	ALL AIR POLLUTION
High	99.00	0.08	**99.08**
Upper middle	44.00	28.00	**72.00**
Lower middle	7.00	8.37	**15.37**
Low	1.00	1.30	**2.30**
Global	**150.00**	**37.74**	187.74

Source: World Bank, based on Preker et al. (2016).

cost-benefit framework, these avoided health and social care expenditures are part of the broader benefits of air pollution reduction. For illustrative purposes they are often presented as a percentage of gross domestic product (GDP). Paradoxically, however, these expenditures increase a country's GDP because they are monetary transactions between economic agents, such as patients and doctors, and create demand for the long value chains of products and services. Therefore, the comprehensive wealth accounting approach promoted by the World Bank deducts the cost of air pollution damage to human health from the value of national assets (Lange, Wodon, and Carey 2018).

In addition to treating the diseases related to air pollution, the health sector has important levers with which to prevent those diseases. First, it can reduce air pollution–related mortality and morbidity through the collection, analysis, and dissemination of health knowledge and information that help build public support for air quality and climate action. Second, sector managers can reduce emissions from facilities. Analysis from the United States finds that health care facilities were responsible for significant fractions of national air pollution, including 10 percent of smog formation, 9 percent of respiratory disease from particulate matter, and 12 percent of acid rain formation (Eckelman 2016).

COST-BENEFIT AND COST-EFFECTIVENESS OF IMPROVING AIR QUALITY

Economic rationale suggests focusing scarce resources on policies and projects that yield the highest surplus of social benefits over social costs.[3] For example, the World Bank (Guigale, Fretes-Cibils, Newman 2007) estimated the economic viability of a series of potential air quality improvement measures in Peru and found that retrofitting existing sources with end-of-pipe particulate filters is the lowest-cost measure with the highest benefit-to-cost ratio, followed by enhanced vehicle inspection and maintenance and end-of-pipe industrial pollution controls. Air quality improvement through fuel switching was relatively costly to achieve, with a lower ratio of benefits to costs (see figure 2.4).

Many constituencies find it too controversial to put a price on the life or health of their citizens and use cost-effectiveness rather than cost-benefit techniques to prioritize policy interventions. Policy makers begin by determining the air quality concentration and exposure objectives that are safe for health and ecosystems. They then call on experts to conduct source-apportionment studies, build models of atmospheric chemistry and pollution dispersion, and identify the sources responsible for exposure of large and the most vulnerable populations to low air quality. Once the key sources polluting an exposed population are established, the policy interventions can focus on those sources and find ways to reduce their emissions of air pollutants and their precursors to achieve the desired level of air quality at the lowest overall cost to society. Cost-effectiveness is commonly applied in Organisation for Economic Co-operation and Development economies and increasingly in developing countries to determine the programs and policies that can attain air quality standards previously established, often within the cost-benefit framework.

The rest of this report provides guidelines on how to harness synergies and manage trade-offs between efforts to reduce air pollution and efforts to mitigate climate change through an integrated policy approach. Chapter 3 explores

FIGURE 2.4

Benefit/cost ratios for air quality measures for Lima-Callao in Peru

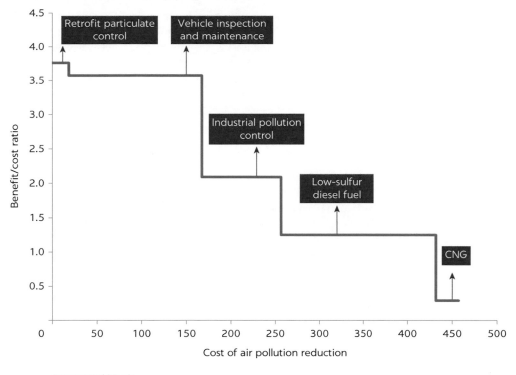

Source: World Bank.
Note: CNG = compressed natural gas.

which polluting substances cause air pollution and climate change at the same time, and which gases can be harmful to health but cool the earth's climate. Chapter 4 identifies the major emissions sources of GHGs and air pollutants and analyzes how they overlap and differ. Chapter 5 examines which abatement measures applied to these sources reduce both air pollutants and climate warmers, and which reduce one while increasing emissions of the other. Chapter 6, which concludes this report, is devoted to the integrated design of policy instruments to leverage synergies and manage trade-offs between measures for air quality and measures for climate change mitigation.

NOTES

1. The short-lived climate pollutant with the longest lifetime in the atmosphere (up to 15 years) is methane, which has a powerful climate-forcing effect (historically, methane is the second-most responsible for global warming after CO_2), while at the same time being a precursor of ground-level ozone.
2. The study includes the following conditions: lower respiratory infections, upper respiratory infections and otitis, perinatal conditions, congenital anomalies, malnutrition, childhood-cluster diseases, cancers, cardiovascular diseases, chronic obstructive pulmonary disease, and asthma.
3. The benefit/cost (B/C) ratio is sometimes used, but it is misleading without additional information about the scale of net benefits. Sprinkling public funds and institutional capacity on multiple small projects with high B/C ratios but small absolute benefits is likely to lead to higher transaction costs and poorer air quality.

REFERENCES

Amann, Markus, M. Holland, R. Maas, T. Vandyck, and B. Saveyn. 2017. *Costs, Benefits and Economic Impacts of the EU Clean Air Strategy and Their Implications on Innovation and Competitiveness.* Laxenburg, Austria: International Institute for Applied Systems Analysis (IIASA).

Chile, Ministerio del Medio Ambiente. 2013. *Primer Reporte del Estado del Medio Ambiente.* Santiago, Chile: Ministerio del Medio Ambiente.

Commission of the European Communities. 2005. "Impact Assessment, Annex to the Communication on Thematic Strategy on Air Pollution and Directive on Ambient Air Quality and Cleaner Air for Europe SEC 1133." Commission Staff Working Paper, Brussels. https://ec.europa.eu/environment/archives/cafe/pdf/ia_report_en050921_final.pdf.

Eckelman, M. 2016. "Environmental Impacts of the U.S. Health Care System and Effects on Public Health." *PLoS One* 11 (6): e0157014.

Guigale, Marcelo M., Vincente Fretes-Cibils, John L. Newman. 2007. *An Opportunity for a Different Peru: Prosperous, Equitable, and Governable.* Washington, DC: World Bank. https://openknowledge.worldbank.org/handle/10986/6633. License: CC BY 3.0 IGO.

Kiesewetter, Gregor, and Markus Amann. 2014. "Urban $PM_{2.5}$ levels under the EU Clean Air Policy Package." TSAP Report #12. https://pure.iiasa.ac.at/id/eprint/11152/1/XO-14-073.pdf.

Lange, Glenn-Marie, Quentin Wodon, and Kevin Carey. 2018. *The Changing Wealth of Nations 2018: Building a Sustainable Future.* Washington, DC: World Bank. https://openknowledge.worldbank.org/handle/10986/29001.

Pigou, A. C. 1920. *The Economics of Welfare.* London: Macmillan.

Preker, Alexander S., Olusoji O. Adeyi, Marisa Gil Lapetra, Diane-Charlotte Simon, and Eric Keuffel. 2016. "Health Care Expenditures Associated with Pollution: Exploratory Methods and Findings." *Annals of Global Health* 82 (5): 711–21. https://annalsofglobalhealth.org/articles/abstract/10.1016/j.aogh.2016.12.003/.

Public Health England. 2018. *Estimation of Costs to the NHS and Social Care Due to the Health Impacts of Air Pollution.* London: Public Health England.

3 Air Pollutants and Greenhouse Gases

INTRODUCTION

This chapter reviews the current state of knowledge about atmospheric chemistry and identifies the extent to which the same gases and particles cause local air pollution and global warming. Air pollution and climate change are triggered by the whirling soup of various chemicals in the air interacting with one another according to the complex rules of dynamic atmospheric chemistry. An understanding of the relations between different pollutants in the atmosphere is the basis for the design of integrated policy packages to address both air pollution and climate problems effectively and cost-effectively (Melamed, Schmale, and von Schneidemesser 2016).

Among the many gases emitted from anthropogenic activities, most warm but some cool the earth's climate. In the technical lingo of climate science, these gases have, respectively, positive or negative "radiative forcing." Relative to the period of industrial revolution, total (net) anthropogenic radiative forcing is strongly positive (see figure 3.1). Virtually all well-mixed, long-lived greenhouse gases (GHGs) warm the climate, whereas the radiative forcing of short-lived gases and aerosols is both positive and negative, with a total net cooling effect. Among air pollutants, sulfur dioxide and, to a lesser extent, nitrogen oxides and ammonia (all also precursors to secondary air pollutants) have a cooling effect. On the other hand, methane, volatile organic compounds, carbon monoxide, and, to a lesser extent, black carbon (BC) show the warming effect. On aggregate, the cooling effect of air pollutants outweighs their warming effect. Methane is an exception, as a second-most-forceful greenhouse gas and, at the same time, a precursor of ground-level ozone (O_3).

One way to illustrate interactions between the local and global impacts of different gases is to follow the precursors of major air pollutants and check their radiative forcing. Figure 3.2, adapted from the IPCC 5th Assessment Report (Myhre et al. 2013) should be read from left to right. If a policy maker wants to reduce ozone pollution or particulate matter concentration (left column), the policies need to target their precursors (listed in the middle column). Reducing emissions of these precursors can have a cooling or warming effect on climate,

FIGURE 3.1

Components of radiative forcing by gases and aerosols

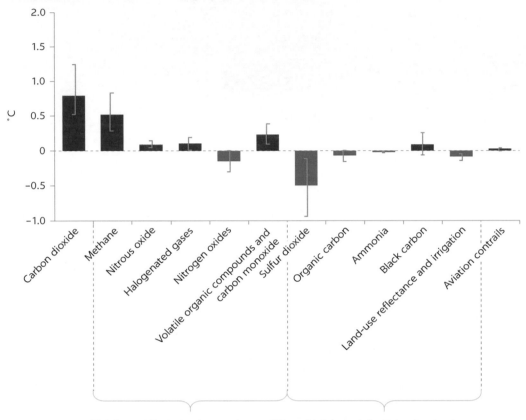

Mainly contribute to changes in non-CO$_2$ greenhouse gases

Mainly contribute to changes in anthropogenic aerosols

Source: IPCC 2021.

Note: The figure shows temperature changes from individual components of human influence: emissions of greenhouse gases, aerosols, and their precursors; land-use changes (land-use reflectance and irrigation); and aviation contrails. Whiskers show very likely ranges. Estimates account for both direct emissions into the atmosphere and their effect, if any, on other climate drivers. For aerosols, both direct effects (through radiation) and indirect effects (through interactions with clouds) are considered (IPCC 2021, p. 7). °C = degrees Celsius; CO$_2$ = carbon dioxide.

or both (right column). For example, controls to reduce pollution by fine particulate matter (PM$_{2.5}$) require reducing emissions of precursors such as black carbon and sulfur dioxide. BC emissions lead to warming the climate (indicating that its abatement is a win-win between air pollution and climate change). Emissions of sulfur dioxide lead to cooling the climate (indicating that its abatement involves trade-offs between air pollution and climate change).

Gases that simultaneously cause air pollution and climate change are as common as those that alleviate one of these environmental challenges while aggravating the other. At least five gases (BC, ground-level ozone, methane, carbon monoxide, and volatile organic compounds) cause adverse health effects of air pollution and warm the planet at the same time. Similar numbers of gases and chemicals are harmful air pollutants but mitigate climate change by cooling the earth (organic carbon [OC], sulfur dioxide [SO$_2$], nitrogen oxides [NO$_x$], ammonia [NH$_3$], and secondary inorganic aerosols), or are neutral to the climate (heavy metals and toxic chemicals). Unspecified particulate matter (other than BC or organic carbon) is a major air pollutant with ambiguous radiative forcing,

FIGURE 3.2

Impact of emission of key primary air pollutants on the most harmful secondary air pollutants and climate

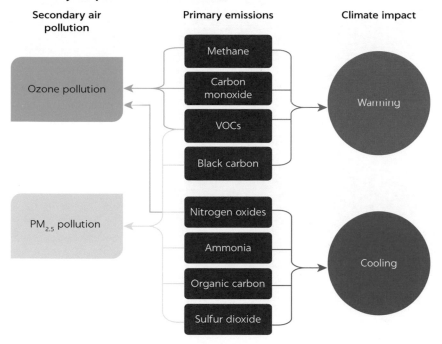

Source: Adaptation based on Myhre et al. 2013.
Note: PM$_{2.5}$ = particulate matter two-and-one-half microns or less in width; VOCs = volatile organic compounds.

depending on circumstances. Four major long-lived GHGs are strong climate warmers but are not (at least directly) harmful to human health. Table 3.1 presents co-benefits and trade-offs between air and climate pollutants.

Local pollutants and their precursors that contribute to climate warming are often referred to as near-term climate forcers (Fiore, Naik, and Leibensperger 2015; Myhre et al. 2013), short-lived climate forcers (SLCFs) (IPCC 2013), or short-lived climate pollutants (SLCPs) (CCAC 2019) because their typical lifetimes are shorter relative to the main GHGs such as carbon dioxide (CO$_2$) (UNEP and WMO 2011). The combined historical radiative forcing of SLCPs is similar to many of the well-mixed long-lived GHGs, such as CO$_2$ (Myhre et al. 2013; IPCC 2021), although the impact of current flow is more temporary. Rogelj et al. (2014) show that the short- and long-term climate effects of many SLCPs become smaller in scenarios that keep warming to less than 2°C relative to preindustrial levels. The SLCPs warm most strongly close to where they are emitted except methane, the atmospheric lifetime of which is around 12 years. Thus, the emission of SLCPs can lead to regional differences in global warming, and their reduction can lead to differential cooling benefits. This difference can be important, for example, near the poles where the rate of warming is much greater than the global average and tipping points are critically dependent on the rate of warming. The local impact of BC is stronger when it is deposited on snow and ice, darkening the surface, increasing melt rates, and further exacerbating heat absorption by the earth's surface (Bond et al. 2013; US EPA 2012; World Bank and ICCI 2013). All SLCPs except hydrofluorocarbons are also causing air pollution. Table 3.2 dives

TABLE 3.1 Impact of pollutants on local human health (through air pollution) and climate change

POLLUTANT	LOCAL HEALTH IMPACT	CLIMATE IMPACT	CO-BENEFITS OR TRADE-OFFS BETWEEN AIR POLLUTION AND CLIMATE CHANGE
1. Black carbon (BC)—component of PM$_{2.5}$	Harmful	Warming	Synergy between air pollution and climate (short-lived climate pollutants)
2. Ground-level ozone (O$_3$)	Harmful	Warming	
3. Methane (CH$_4$)	Harmful	Warming	
4. Carbon monoxide (CO)	Harmful	Warming	
5. Volatile organic compounds	Harmful	Warming	
6. Organic carbon (OC)	Harmful	Cooling	Trade-offs between air pollution and climate mitigation
7. Sulfur dioxide (SO$_2$)	Harmful	Cooling	
8. Nitrogen oxides (NO$_X$)	Harmful	Cooling	
9. Ammonia (NH$_3$)	Harmful	Cooling	
10. Secondary inorganic aerosols	Harmful	Cooling	
11. Heavy metals, benzo[a]pyrene, dioxins, and so on	Harmful	Neutral	
12. Carbon dioxide (CO$_2$)	Neutral	Warming	Long-term climate forcers, neutral for air quality
13. Chlorofluorocarbons (CFCs)	Neutral	Warming	
14. Hydrofluorocarbons (HFCs)	Neutral	Warming	
15. Nitrous oxide (N$_2$O)	Neutral	Warming	

Source: World Bank.
Note: PM$_{2.5}$ = particulate matter two-and-one-half microns or less in width.

deeper into the climate and local air pollution effects of the SLCPs. The focus on SLCPs led to establishment of the Climate and Clean Air Coalition to Reduce Short-Lived Climate Pollutants.

Fine particulates (PM$_{2.5}$ and PM$_{10}$) warm or cool the climate depending on their chemical composition, but in most cases the overall warming effect dominates, especially with a high share of BC in atmospheric aerosols. The impact of total PM$_{2.5}$ and PM$_{10}$ on local air pollution and climate change differs across sources, locations, and technologies. For example, biomass burning simultaneously emits two types of particulates with opposite effects on climate: cooling OC and warming BC. Although the direct warming or cooling of particle pollution is well understood, there are several indirect effects of particle pollution on the climate system (for example, cloud properties and distribution, snow and ice deposition) with significantly greater uncertainty (Myhre et al. 2013). Some estimates suggest significant net warming potential (Boucher et al. 2016). Climate benefits depend on how complete combustion is (that is, the ratio of BC to OC) and the location relative to surface brightness. As discussed above, stronger radiative forcing is associated with emissions affecting light snow or ice, because even reflective aerosol is darker than fresh snow (US EPA 2012; World Bank and ICCI 2013). A weaker warming or even a cooling effect occurs when particles fall on darker surfaces, such as oceans or tropical forests (Bond et al. 2013).

Among aerosols, the sulfates, mostly from fossil fuel use, have a dominant role in climate-cooling effects, whereas BC, mostly from biofuel use, is the key source of positive radiative forcing (table 3.3). Several air pollutants interact in the atmosphere with fine particles and form secondary aerosols. Some particles, such as BC, OC, and sea salt, are primary aerosols. This is also the case with windblown mineral dust from desert areas, unpaved roads, and soil disturbance

TABLE 3.2 Short-lived climate pollutants (SLCPs) are air pollutants and climate forcers

Black carbon (BC): Lifetime in atmosphere is days to weeks

Sources: BC, or soot, is one of the components of particulate matter (PM) comprising, on average, 10 percent of $PM_{2.5}$ by mass in ambient air (Briggs and Long 2016; Chen et al. 2014; Chow et al. 2002; Chow et al. 2011; Christoforou et al. 2000), but it can reach up to 30 percent depending on the location and sources (CCAC 2018; Gramsch et al. 2014). A large share of BC's warming effect is local, transmitted through changing patterns of clouds and rain and covering the snow and ice, hence decreasing the earth's ability to reflect solar warming radiation, thereby absorbing heat and accelerating melt. BC is formed by the incomplete combustion of fossil fuels, wood, and other (mainly bio-) fuels and comprises between 10 percent and 90 percent of primary emissions of $PM_{2.5}$, depending on the source. The majority of global BC emissions come from household cooking, heating, other inefficient small burners and stoves, brick kilns and coke ovens used in the developing world (Rajarathnam et al. 2014), open biomass and waste burning, flaring of gas associated with oil extraction, and diesel engines. Diesel vehicles (pre–Euro 6/VI emissions standards) also emit soot, which is almost pure BC. Power stations using heavy oil or coal also emit BC, especially old, inefficient units without pollution-control equipment (dust filters, sulfur scrubbers). As a result, Africa, Asia, and Latin America contribute approximately 88 percent of global BC emissions (CCAC 2018).

Climate impact	*Air quality impact*
Although it does not trap the heat in the atmosphere, BC makes the earth's surface darker and prone to absorbing more heat and accelerated melting when deposited on snow and ice (Myhre et al. 2013). BC affects the climate globally, but mainly regionally by changing the patterns of clouds and rain. BC is responsible for a significant proportion of the radiative forcing to date (Bond et al. 2013; Boucher et al. 2016; Collins et al. 2013; Myhre et al. 2013; Ramanathan and Carmichael 2008), but the warming effect is not long lasting. BC could sometimes also lead to some cooling effect; however, its interactions with clouds are not fully understood (CCAC 2018; Myhre et al. 2013). The local warming impact of BC is larger when it is deposited on snow or ice. The most recent IPCCs 6th Assessment Report found that the effective radiative forcing attributed to black carbon is rather small: +0.08 [with the uncertainty range from 0.00 to 0.18] Watts per square meter (W/m²), substantially smaller than previously assessed (Arias et al. 2021).	Although the specific health impacts of BC are hard to establish, BC forms an important part of $PM_{2.5}$. Thus, measures to reduce BC will necessarily also reduce $PM_{2.5}$, for which the positive health effects are well established and include respiratory and cardiovascular diseases, cancer, and even birth defects (Janssen et al. 2012; US EPA 2012). Emissions reduction of BC, brown carbon, and organic carbon from residential solid biomass combustion will have large public health benefits when emissions can be reduced by 80 percent or more (Anenberg et al. 2012).

Ground-level ozone (O_3): Lifetime in atmosphere is hours to weeks

Sources: Ozone is not emitted directly in significant quantities but rather is formed in the air when the emission of VOCs, NO_x, methane, and CO are combined in the presence of sunlight. Therefore, ozone is the main culprit behind the "summer smog" in many urban areas. The main sources of precursor emissions include vehicles, power plants, and industrial activities (for NO_x); fugitive emissions from dispersed consumption of chemicals, such as paints, solvents, aerosol sprays, cleansers, dry cleaning of clothing, and so on (for VOCs); and agriculture, municipal utilities, and fossil fuel extraction (such coal mines) and transport (gas networks) (for methane). Close to the earth's surface, ozone is an air pollutant, but in the stratosphere, it protects the life forms on the earth by absorbing 97 percent to 99 percent of the sun's ultraviolet radiation.

Climate impact	*Air quality impact*
Ground-level ozone is considered a short-lived climate pollutant. Arias et al. 2021 assessed that the total ozone effective radiative forcing for total (stratospheric and ground-level) ozone is 0.47 [uncertainty range 0.24 to 0.71] Wm², slightly more than earlier estimates by Fiore, Naik, and Leibensperger 2015). Myhre et al. (2013) and Shindell et al. (2013), but significantly larger than black carbon.	Ozone is a powerful respiratory irritant responsible for between 245,000 (Cohen et al. 2017) and 1 million (CCAC 2022) premature deaths and more than 4 million disability-adjusted life years attributable to chronic obstructive pulmonary disease each year (Cohen et al. 2017). Ground-level ozone is also harmful to plants, including crop yields (Hartmann et al. 2013).

Methane (CH_4): Lifetime in the atmosphere is about 10 years

Sources: Emitted mainly from agricultural activities (for example, raising livestock, rice cultivation), landfills, solid waste and wastewater treatment facilities, and coal mines, as well as from the production rigs and distribution of oil and natural gas.

Climate impact	*Air quality impact*
Methane is also a powerful GHG, second after CO_2, responsible for about 20 percent of global warming so far, despite its relatively short atmospheric lifetime (about 10 years) compared with CO_2. Arias et al. 2021 assessed that the effective radiative forcing due to methane emissions is 1.21 [uncertainty range 0.90 to 1.51] W/m², by far the highest of all SLCPs.	Methane is not a directly harmful air pollutant but acts as a precursor of ground-level ozone (see section on ground-level ozone).

continued

TABLE 3.2, *Continued*

Volatile organic compounds (VOCs)	
Sources: VOCs are emitted from fugitive sources related to the dispersed consumption of chemicals such as paints, solvents, aerosol sprays, cleansers, dry cleaning of clothing, and so on, as well as fuel combustion (mainly by vehicles) and from biomass burning, including cook stoves. Under sunlight, VOCs react with NO_x emitted mainly from vehicles, power plants, and industrial activities to form ground-level ozone, which in turn helps the formation of fine particulates.	
Climate impact	**Air quality impact**
VOCs have an indirect warming effect (through ozone formation); according to Arias et al. 2021 VOCs' warming effect is small compared with the warming effect of methane but jointly with CO larger than any other SLCP.	VOCs are harmful to health both directly and as precursors of ground-level ozone and fine particles.

Carbon monoxide (CO)	
Sources: CO is formed during incomplete combustion of fossil fuels.	
Climate impact	**Air quality impact**
The radiative forcing, though small and indirect, is positive because CO also participates in ground-level ozone formation.	CO is toxic when inhaled directly and has chronic health affects in lower concentrations, for example, through cardiovascular risk.

Source: World Bank based on multiple sources.
Note: BC = black carbon; CO = carbon monoxide; CO_2 = carbon dioxide; GHG = greenhouse gas; NO_x = nitrogen oxides; O_3 = ozone; $PM_{2.5}$ = particulate matter two-and-one-half microns or less in width; SLCP = short-lived climate pollutant; VOCs = volatile organic compounds.

from construction. Except for BC, other primary organic aerosols tend to reflect more sunlight than they absorb, cooling the climate. Pollutants such as SO_2, NO_x, and ammonia (see table 3.3) interact in the atmosphere to form secondary inorganic aerosols, composed of a mixture of sulfates, nitrates, ammonium, and fine particles (including $PM_{2.5}$). The primary particles BC, sea salt, mineral dust, and so on also contribute to aerosols. They show cooling effect by reflecting and scattering incoming sunlight, although they can also force local or regional perturbations to climate (CCAC 2018; Myhre et al. 2013; Arias et al. 2021). Although the cooling properties of secondary inorganic aerosols are well established, their effect is rather short-lived compared with the long-lived GHGs, such as CO_2, which will continue warming the climate after 2050 (Sokan-Adeaga et al. 2019). The fifth IPCC Assessment Report noted that "If rapid reductions in sulfate aerosol are undertaken for improving air quality or as part of decreasing fossil-fuel CO_2 emissions, then there is medium confidence that this could lead to rapid near-term warming" (IPCC 2013, 81). The sixth IPCC Assessment Report stipulates that air-polluting aerosols have so far canceled 27 percent of the total global warming resulting from GHG emissions (figure 3.3). The best estimates of effective radiative forcing (ERF) attributed to sulfur dioxide (SO_2) and CH_4 emissions are substantially greater than in the fifth Assessment Report, while that of black carbon is substantially reduced. The magnitude of uncertainty in the ERF due to black carbon emissions has also been reduced relative to AR5 (the fifth Assessment Report) (Arias et al. 2021).

Kloster et al. (2010) estimates that reducing aerosol emissions worldwide by 2030 could increase the equilibrium temperature by 0.96°C. Shindell et al. (2012) explores the climate benefits of a selective SLCP strategy that maximizes the reduction of warming SLCPs while minimizing the reduction of cooling SLCPs. Such a strategy could reduce projected warming by close to 0.5°C in 2030, but at the cost of an air pollution penalty.

TABLE 3.3 Local pollutants (aerosol precursors) that are climate coolants

Sulfur dioxide (SO_2): Lifetime in the atmosphere is days to weeks
Sources: SO_2 emissions come primarily from the burning of sulfur-containing fossil fuels, mainly by power plants and other industrial facilities. Smaller sources of SO_2 emissions include industrial processes such as extracting metal from ore; natural sources such as volcanoes; and locomotives, ships, and other vehicles and heavy equipment that burn fuel with a high sulfur content (US EPA 2021b).

Climate impact	*Air quality impact*
SO_2 reacts with fine particles and forms complex sulfate aerosols in the atmosphere, which reflect and scatter the heat energy of sunlight and cool the planet.	SO_2 is directly harmful to human health, damages fragile ecosystems (for example, pine forests), destroys structures when dissolved in rain or mist as acids, and changes the chemistry of rainwater, leading to acidification of ecosystems. As a precursor for secondary $PM_{2.5}$ it causes important health damage. Recent research suggests that secondary $PM_{2.5}$ that originate from sulfate aerosols from fossil fuel combustion are particularly harmful (World Bank 2021).

Nitrogen oxides (NO_x)
Sources: NO_x gases form when fuel is burned through oxidation of nitrogen. There are three primary sources of NO_x formation in combustion process. First, thermal NO_x formation (from nitrogen found in air) is highly temperature dependent and is recognized as the most relevant source when combusting natural gas. Second, oxidation of nitrogen contained in the fuel itself (fuel NO_x) tends to dominate during the combustion of fuels, such as coal, which have a significant nitrogen content. The third source, "prompt NO_x" is also formed from the nitrogen contained in atmospheric air and is considered the smallest of the three. NO_x pollution is emitted by automobiles, trucks, and various nonroad vehicles (for example, construction equipment, boats, and so on) as well as by industrial sources such as power plants, industrial boilers, cement kilns, and turbines (US EPA 2021a). The temperature of combustion is important in NO_x formation. Combustion at temperatures well below 1,300°C forms much smaller concentrations of NO_x. NO_x emissions also come from fertilizer use.

Climate impact	*Air quality impact*
The impact on climate change is similar to that of SO_2. NO_x reacts with other gases (for example, ammonia) and forms nitrate aerosols, which act as climate coolants. NO_x gases also cause climate warming because of their impact on methane lifetime and aerosol formation (Myhre et al. 2013), but the net impact on climate change is cooling.	NO_x are poisonous, highly reactive gases directly harmful to health and contribute to "acid rain," which causes eutrophication, harms plants, and reduces carbon sequestration of fragile ecosystems. However, their major impact on air pollution is as precursors of aerosols and fine particles (especially $PM_{2.5}$), as well as ground-level ozone.

Ammonia (NH_3)
Sources: The main sources are ammonium carbonate and inorganic fertilizers used in agriculture and monogastric animal production.

Climate impact	*Air quality impact*
NH_3 has a mild cooling effect on climate, but global warming is increasing NH_3 emissions from agriculture (Schauberger et al. 2018).	NH_3 has an indirect harmful impact on air pollution as an important precursor of fine particles (especially $PM_{2.5}$).

Source: World Bank based on multiple sources.
Note: NO_x = nitrogen oxides; $PM_{2.5}$ = particulate matter two-and-one-half microns or less in width; SO_2 = sulfur dioxide.

The most potent GHGs do not cause immediate local health damage. CO_2 is the most important GHG but is harmless to local health. It has a complex indirect impact on health through the long-term impacts of climate change (WHO 2018), but locally it can be harmful to health only in exceptional circumstances; for example, in high concentrations in the basements of indoor spaces, CO_2 can displace oxygen in the air as a heavier gas. Chlorofluorocarbons (CFCs), emitted mainly from aerosols and refrigerators, are long-lived GHGs with relatively low quantities emitted but very strong global warming potential. Their negative health impact is transmitted mainly through the global effect of destroying the stratospheric ozone layer. Hydrofluorocarbons were introduced as an alternative to ozone-damaging CFCs and are also GHGs, though live shorter

FIGURE 3.3

Contributions to warming by sources of human influence

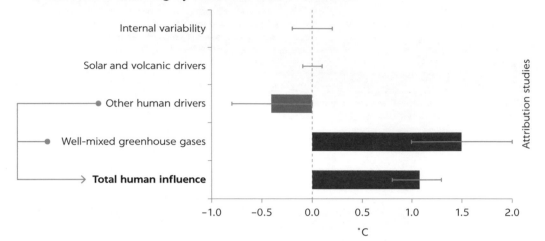

Source: IPCC 2021, p.7.
Note: The figure shows the aggregated contributions to 2010–2019 warming in °C relative to 1850–1900.

than SLCPs. Nitrous oxide (N_2O) is another important GHG emitted by human activities such as fertilizer use and fossil fuel burning. Natural processes in soils and the oceans also release N_2O. It can be a health hazard in large indoor concentrations, but rarely in ambient air.

Several harmful local air pollutants are neutral with respect to climate. Heavy metals include lead, chromium, mercury, cadmium, arsenic, copper, manganese, nickel, zinc, and silver. Some are carcinogenic and some result in neurological deficits. Major sources include industrial emissions and vehicle exhaust, mostly in particulate form.

Pursuing ambitious climate-mitigation policies can increase air pollution in certain airsheds. Some climate-mitigation pathways, especially those consistent with the 1.5°C temperature goal, can rely heavily on the steep increase of demand for biofuels. A strong increase of the use of biomass for energy generation, especially domestic heating and cooking and biofuels use in transport, can result in increased emissions of $PM_{2.5}$ and its precursors (Rogelj, Pop, et al. 2018; Rogelj, Shindell, et al. 2018).

This chapter demonstrates that an integrated approach requires understanding that relatively few gases produce a double environmental dividend, that is, a dividend that helps mitigate both air pollution and climate change. Such synergy is present for SLCPs such as BC, methane, and ground-level ozone with its precursors. However, reducing emissions of sulfate and nitrate aerosol precursors, in particular SO_2 and NO_x, mitigates adverse health effects and damage to some fragile ecosystems but accelerates global warming. Reduction of population exposure to toxic and carcinogenic air pollutants does not help climate. The longest-lived and well-mixed GHGs, including, most important, CO_2, are not contributing to air pollution. These opposite effects of non-BC aerosols on air pollution and climate change need to be recognized and reflected in policy design. Rapid reduction of air pollutants is essential to saving 7 million lives lost to air pollution globally every year. **If air quality programs reduce emissions of climate coolants, it just implies that more ambitious climate action is needed**

than previously anticipated to stay within the 2°C objectives of the Paris Agreement. Chapter 4 discusses the extent to which the major sources of local pollution and GHGs overlap. Chapter 5 turns to choices of abatement measures that maximize synergies and manage trade-offs between goals relating to air quality on the one hand, and goals relating to mitigating climate change on the other hand.

REFERENCES

Anenberg, Susan C., Joel Schwartz, Drew Shindell, Markus Amann, Greg Faluvegi, Zbigniew Klimont, Greet Janssens-Maenhout, et al. 2012. "Global Air Quality and Health Co-Benefits of Mitigating Near-Term Climate Change through Methane and Black Carbon Emission Controls." *Environmental Health Perspectives* 120 (6): 831–39. doi:10.1289/ehp.1104301.

Arias, P.A., N. Bellouin, E. Coppola, R. G. Jones, G. Krinner, J. Marotzke, V. Naik, M. D. Palmer, G-K Plattner et al. 2021. "Technical Summary." In Climate Change 2021: *The Physical Science Basis. Contribution of Working Group I to the Sixth Assessment Report of the Intergovernmental Panel on Climate Change* [Edited by Masson-Delmotte, V., P. Zhai, A. Pirani, S. L. Connors, C. Péan, S. Berger, N. Caud, Y. Chen, L. Goldfarb, M. I. Gomis, M. Huang, K. Leitzell, E. Lonnoy, J. B. R. Matthews, T. K. Maycock, T. Waterfield, O. Yelekçi, R. Yu, and B. Zhou]. Cambridge University Press, Cambridge, United Kingdom, and New York, United States.

Bond, T. C., S. J. Doherty, D. W. Fahey, P. M. Forster, T. Berntsen, B. J. DeAngelo, M. G. Flanner, et al. 2013. "Bounding the Role of Black Carbon in the Climate System: A Scientific Assessment." *Journal of Geophysical Research: Atmospheres* 118: 5380–552. https://doi.org/10.1002/jgrd.50171.

Boucher, Olivier, Yves Balkanski, Øivind Hodnebrog, Cathrine Lund Myhre, Gunnar Myhre, Johannes Quaas, Bjørn Hallvard Samset, Nick Schutgens, Philip Stier, and Rong Wang. 2016. "Jury Is Still Out on the Radiative Forcing by Black Carbon." *Proceedings of the National Academy of Sciences* 113 (35): E5092–E5093. doi:10.1073/pnas.1607005113.

Briggs, N. L., and C. M. Long. 2016. "Critical Review of Black Carbon and Elemental Carbon Source Apportionment in Europe and the United States." *Atmospheric Environment* 144: 409–27.

CCAC (Climate and Clean Air Coalition). 2018. "2018 Annual Science Update—Black Carbon Briefing Report." CCAC. https://www.ccacoalition.org/en/resources/2018-annual-science-update-black-carbon-briefing-report.

CCAC (Climate and Clean Air Coalition). 2019. "Definitions of Short-Lived Pollutants," CCAC (accessed November 24, 2019), https://ccacoalition.org/en/content/short-lived-climate-pollutants-slcps.

CCAC. 2022. "Tropospheric Ozone," CCAC (accessed April 20, 2022) https://www.ccacoalition.org/en/slcps/tropospheric-ozone.

CCAC (Climate and Clean Air Coalition) and UNEP (United Nations Environment Programme). 2019. *Air Pollution in Asia and the Pacific: Science-Based Solutions.* Bangkok: Asia Pacific Clean Air Partnership.

Chen, X., Z. Zhang, G. Engling, R. Zhang, J. Tao, M. Lin, X. Sang, C. H. Chan, S. Li, and Y. Li. 2014. "Characterization of Fine Particulate Black Carbon in Guangzhou, a Megacity in South China." *Atmospheric Pollution Research* 5: 361–70.

Chow, J. C., J. G. Watson, S. A. Edgerton, and E. Vega. 2002. "Chemical Composition of $PM_{2.5}$ and PM_{10} in Mexico City during Winter 1997." *Science of the Total Environment* 287: 177–201.

Chow, J. C., J. G. Watson, D. H. Lowenthal, L. W. Antony Chen, and N. Motallebi. 2011. "$PM_{2.5}$ Source Profiles for Black and Organic Carbon Emission Inventories." *Atmospheric Environment* 45: 5407–14.

Christoforou, C., L. Salmon, M. Hannigan, P. Solomon, and G. R. Cass. 2000. "Trends in Fine Particle Concentration and Chemical Composition in Southern California." *Journal of the Air and Waste Management Association* 50: 43–53.

Cohen, Aaron J., Michael Brauer, Richard Burnett, H. Ross Anderson, Joseph Frostad, Kara Estep, Kalpana Balakrishnan, et al. 2017. "Estimates and 25-Year Trends of the Global Burden of Disease Attributable to Ambient Air Pollution: An Analysis of Data from the Global Burden of Diseases Study 2015." *Lancet* 389 (10082): 1907–18. https://doi.org/10.1016/S0140 -6736(17)30505-6.

Collins, W. J., M. M. Fry, H. Yu, J. S. Fuglestvedt, D. T. Shindell, and J. J. West. 2013. "Global and Regional Temperature-Change Potentials for Near-Term Climate Forcers." *Atmospheric Chemistry and Physics* 13 (5): 2471–85. doi:10.5194/acp- 13-2471-2013.

Fiore, A. M., V. Naik, and E. M. Leibensperger. 2015. "Air Quality and Climate Connections." *Journal of the Air and Waste Management Association* 65 (6): 645–85. doi:10.1080/10962247 .2015.1040526.

Gramsch, E., D. Caceres, P. Oyola, F. Reyes, Y. Vasquez, M. A. Rubio, and G. Sanchez. 2014. "Influence of Surface and Subsidence Thermal Inversion on $PM_{2.5}$ and Black Carbon Concentration." *Atmospheric Environment* 98: 290–98.

Hartmann, D. L., A. M. G. Klein Tank, M. Rusticucci, L. V. Alexander, S. Brönnimann, Y. Charabi, F. J. Dentener, et al. 2013. "Observations: Atmosphere and Surface." In *Climate Change 2013: The Physical Science Basis. Contribution of Working Group I to the Fifth Assessment Report of the Intergovernmental Panel on Climate Change*, edited by Thomas F. Stocker, Dahe Qin, Gian-Kasper Plattner, Melinda M. B. Tignor, Simon K. Allen, Judith Boschung, Alexander Nauels, Yu Xia, Vincent Bex, and Pauline M. Midgley. Cambridge, UK, and New York, NY: Cambridge University Press.

IPCC (Intergovernmental Panel on Climate Change). 2013. *Climate Change 2013: The Physical Science Basis. Contribution of Working Group I to the Fifth Assessment Report of the Intergovernmental Panel on Climate Change*, edited by T. F. Stocker, D. Qin, G.-K. Plattner, M. Tignor, S. K. Allen, J. Boschung, A. Nauels, Y. Xia, V. Bex, and P. M. Midgley. Cambridge, UK, and New York, NY: Cambridge University Press. www.ipcc.ch/report/ar5/wg1.

IPCC. 2021. "Summary for Policymakers." In *Climate Change 2021: The Physical Science Basis. Contribution of Working Group I to the Sixth Assessment Report of the Intergovernmental Panel on Climate Change* [Edited by Masson-Delmotte, V., P. Zhai, A. Pirani, S. L. Connors, C. Péan, S. Berger, N. Caud, Y. Chen, L. Goldfarb, M. I. Gomis, M. Huang, K. Leitzell, E. Lonnoy, J. B. R. Matthews, T. K. Maycock, T. Waterfield, O. Yelekçi, R. Yu, and B. Zhou (eds.)].

Janssen, Nicole A. H., Miriam E. Gerlofs-Nijland, Timo Lanki, Raimo O. Salonen, Flemming Cassee, Gerard Hoek, Paul Fischer, Bert Brunekreef, and Michal Krzyzanowski. 2012. *Health Effects of Black Carbon*. Geneva: World Health Organization.

Kloster, Silvia, Frank Dentener, Johann Feichter, Frank Raes, Ulrike Lohmann, Erich Roeckner, and Irene Fischer-Bruns. 2010. "A GCM Study of Future Climate Response to Aerosol Pollution Reductions." *Climate Dynamics* 34: 1177. https://doi.org/10.1007/s00382-009 -0573-0.

Melamed, M. L., J. Schmale, and E. von Schneidemesser. 2016. "Sustainable Policy: Key Considerations for Air Quality and Climate Change." *Current Opinion in Environmental Sustainability* 23: 85–91. https://doi.org/10.1016/j.cosust.2016.12.003.

Myhre, G., D. Shindell, F. M. Bréon, W. Collins, J. Fuglestvedt, J. Huang, D. Koch, et al. 2013. "Anthropogenic and Natural Radiative Forcing." In *Climate Change 2013: The Physical Science Basis. Contribution of Working Group I to the Fifth Assessment Report of the Intergovernmental Panel on Climate Change*, edited by T. F. Stocker, D. Qin, G.-K. Plattner, M. Tignor, S. K. Allen, J. Boschung, A. Nauels, Y. Xia, V. Bex, and P. M. Midgley. Cambridge, UK, and New York, NY: Cambridge University Press.

Rajarathnam, U., V. Athalye, S. Ragavan, S. Maithel, D. Lalchandani, S. Kumar, E. Baum, C. Weyant, and T. Bond. 2014. "Assessment of Air Pollutant Emissions from Brick Kilns." *Atmospheric Environment* 98: 549–33.

Ramanathan, V., and G. Carmichael. 2008. "Global and Regional Climate Change Due to Black Carbon." *Nature Geoscience* 1: 221–27.

Rogelj, J., A. Pop., K. V. Calvin, G. Luderer, J. Emmerling, D. Gernaat, S. Fujimori, et al. 2018. "Scenarios towards Limiting Global Mean Temperature Increase below 1.5 °C." *Nature Climate Change* 8: 325–32. https://doi.org/10.1038/s41558-018-0091-3, with supplementary materials, https://static-content.springer.com/esm/art%3A10.1038%2Fs41558-018-0091-3 /MediaObjects/41558_2018_91_MOESM1_ESM.pdf.

Rogelj, J., M. Schaeffer, M. Meinshausen, D. T. Shindell, W. Hare, Z. Klimont, G. J. M. Velders, M. Amann, and H. J. Schellnhuber. 2014. "Disentangling the Effects of CO_2 and Short-Lived Climate Forcer Mitigation." *Proceedings of the National Academy of Sciences of the United States of America* 111 (46): 16325–30. doi:10.1073/pnas.1415631111.

Rogelj, J., D. Shindell, K. Jiang, S. Fifita, P. Forster, V. Ginzburg, C. Handa, et al. 2018. "Mitigation Pathways Compatible with 1.5°C in the Context of Sustainable Development." In *Global Warming of 1.5°C: An IPCC Special Report on the Impacts of Global Warming of 1.5°C above Pre-Industrial Levels and Related Global Greenhouse Gas Emission Pathways, in the Context of Strengthening the Global Response to the Threat of Climate Change, Sustainable Development, and Efforts to Eradicate Poverty*, edited by V. Masson-Delmotte, P. Zhai, H.-O. Pörtner, D. Roberts, J. Skea, P. R. Shukla, A. Pirani, et al. Geneva: Intergovernmental Panel on Climate Change.

Schauberger, Günther, Martin Piringer, Christian Mikovits, Werner Zollitsch, Stefan J. Hörtenhuber, Johannes Baumgartner, Knut Niebuhr, et al. 2018. "Impact of Global Warming on the Odour and Ammonia Emissions of Livestock Buildings Used for Fattening Pigs." *Biosystems Engineering* 175 (November): 106–14. https://www.sciencedirect.com /science/article/pii/S1537511018305786.

Shindell, D. T., J. C. I. Kuylenstierna, E. Vignati, R. van Dingenen, M. Amann, Z. Klimont, S. C. Anenberg, et al. 2012. "Simultaneously Mitigating Near-Term Climate Change and Improving Human Health and Food Security." *Science* 335: 183–89. doi:10.1126 /science.1210026.

Shindell, D. T., O. Pechony, A. Voulgarakis, G. Faluvegi, L. Nazarenko, J.-F. Lamarque, K. Bowman, et al. 2013. "Interactive Ozone and Methane Chemistry in GISS-E2 Historical and Future Climate Simulations." *Atmospheric Chemistry and Physics* 13: 2653–89.

Sokan-Adeaga, Adewale, Ana Godson Ree, Michael Sokan-Adeaga, Eniola Sokan-Adeaga, and Oseji Ejike. 2019. "Secondary Inorganic Aerosols: Impacts on the Global Climate System and Human Health." *Biodiversity International Journal* 3 (6): 249–59. doi:10.15406 /bij.2019.03.00152.

UNEP (United Nations Environment Programme) and WMO (World Meteorological Organization). 2011. *Integrated Assessment of Black Carbon and Tropospheric Ozone*. Nairobi: UNEP.

US EPA (United States Environmental Protection Agency). 2012. *Report to Congress on Black Carbon*. Washington, DC: US EPA.

US EPA (United States Environmental Protection Agency). 2021a. "Nitrogen Oxides (NO_x) Control Regulations" (accessed July 2021), https://www3.epa.gov/region1/airquality/nox .html.

US EPA (United States Environmental Protection Agency). 2021b. "Sulfur Dioxide (SO_2) Pollution" (accessed July 2021), https://www.epa.gov/so2-pollution/sulfur-dioxide-basics.

World Bank. 2021. *Are All Air Pollution Particles Equal? How Constituents and Sources of Fine Air Pollution Particles ($PM_{2.5}$) Affect Health*. Washington, DC: World Bank.

World Bank and ICCI (International Cryosphere Climate Initiative). 2013. *On Thin Ice: How Cutting Pollution Can Slow Warming and Save Lives*. Washington, DC: World Bank.

4 Emission Sources of Greenhouse Gases and Air Pollutants

INTRODUCTION

This chapter identifies major emission sources of greenhouse gases (GHGs) and air pollutants and explores how they overlap. The measures for reducing emissions from these sources are discussed in chapter 5.

EMISSION SOURCES

Air pollutants and climate forcers are often co-emitted from the same sources. This observation underpins a common argument that addressing air pollution and climate change have bi-directional co-benefits (Agee et al. 2014). Unfortunately, the emission sources that contribute most to air pollution do not always match the sources that contribute most to climate change and vice versa.

Decision-makers need to know where the pollution originates to address problems at their primary source. The key sources of exposure to air pollutants in a polluted airshed, including sources of precursors of these pollutants, can be surprisingly different from what appears to the naked eye. Source-apportionment studies are conducted to find the types of sources (for example, whether the pollutants come from power plants, household boilers, industry, agriculture, or transport) (see figure 4.1). Additional tools, such as pollution dispersion and atmospheric chemistry models, are needed to identify the location and formation of polluting substances in the airshed. For example, the ambient air pollution of Greater Cairo surpasses both the World Health Organization's and the Egyptian national legislation's recommended guideline for clean air by several-fold. The premature deaths of roughly 11,500 people a year in Cairo is associated with ambient air pollution. Much of the pollution, about one-third, comes from motor vehicles (figure 4.1). Another third comes from agricultural and municipal waste burning, making reducing pollution from vehicles and waste burning the priority areas in which to engage.

Emissions that are critical to air pollution sometimes come from sources that also contribute to global warming and vice versa. Table 4.1 matches the main emission sources (in columns) with the polluting substances (rows).

FIGURE 4.1

Source apportionment of PM$_{2.5}$ in Greater Cairo

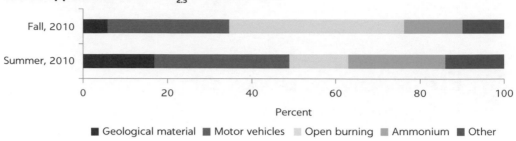

Sources: World Bank 2018.
Note: PM$_{2.5}$ = particulate matter two-and-one-half microns or less in width.

The darker the cell, the larger the share of a pollution source in the total emissions of a substance. Internal combustion vehicles represent the "perfect storm" of air-polluting and climate-warming sources, although chapter 5 discusses some trade-offs related to nitrogen oxides (NO$_x$) abatement from these sources. Waste and agriculture are the major sources of both air pollution and global warming, mainly because of emissions of methane, which is a powerful GHG, albeit with a relatively short lifetime in the atmosphere (about 10 years). Yet, ammonium emitted from agricultural livestock and synthetic fertilizers is a precursor of both black carbon (Updyke, Nguyen, Nizkorodov 2012) and climate-cooling aerosols. Agriculture is a rather small source of the long-lived GHGs such as carbon dioxide.

The overlap between the key sources of long-lived GHG emissions and the key sources of short-lived climate pollutants is only partial (see also Radu et al. 2016). Black carbon is always co-emitted with other particles and gases, although many of them have a cooling effect on the climate (organic carbon, sulfur dioxide [SO$_2$], and NO$_x$, as discussed in chapter 3). Sources that release a high ratio of warming to cooling pollutants are the most promising targets for jointly achieving climate and health benefits in the near term (see table 4.1).

Priority sources for air quality improvement and climate action do not always match well, however. As demonstrated by Lvovsky et al. (2000) for six very different cities, the greatest part of damage by far to local health came from emissions of small households, commercial and industrial boilers and stoves, open burning of municipal or agricultural waste, and from vehicles, rather than from large industries and power plants. In contrast, large industry and power plants co-emit a bulk of GHGs and large volumes of air pollutants, but among the latter, mainly SO$_2$ and NO$_x$, which happen to be climate coolants. The sources responsible for most of the PM$_{2.5}$ concentrations in India include, in order of significance, residential energy use (mainly cooking), transport, and municipal and agricultural waste. The relative contribution of power plants and industry to air pollution varies significantly by province and is often related to small artisanal industries, such as brick kilns rather than large power and industrial plants. (Purohit et al. 2019).

Policy makers who want to quickly prevent diseases and premature deaths associated with air pollution would focus on different sources than policy makers who prioritize low-carbon policies. To quickly lower morbidity and mortality, priorities would be small, low-stack biomass combustion, cooking and heating sources in households, and artisanal industries, as well as burning of

TABLE 4.1 Key emission sources of air pollutants, short-lived climate pollutants, and greenhouse gases

PRIMARY POLLUTANTS	AIR POLLUTION	CLIMATE CHANGE	SOURCES[a] LARGE POWER PLANTS AND INDUSTRY	SMALL ARTISANAL INDUSTRY	HOUSEHOLD HEATING AND COOKING	VEHICLES	WASTE	OPEN BIOMASS BURNING	FERTILIZERS, ANIMAL FARMING	CONSUMPTION GOODS[b]
1. Black carbon (BC)										
2. Methane (CH_4)										
3. Carbon monoxide (CO)										
4. Volatile organic compounds										
5. Unspecified PM_{10}, $PM_{2.5}$										
6. Organic carbon (OC)										
7. Sulfur dioxide (SO_2)										
8. Nitrogen oxides (NO_x)										
9. Ammonia (NH_3)										
10. Heavy metals										
11. Carbon dioxide (CO_2)										
12. Chlorofluorocarbons (CFCs)										
13. Hydrofluorocarbons (HFCs)										
14. Nitrous oxide (N_2O)										

Type of impact—Color coding: red = negative impact; yellow = negative and positive impact; gray = no impact; and green = positive impact.

Air pollution | Climate change | Major source | Secondary source | Minor source

Source: World Bank.

Note: PM_{10} = particulate matter 10 microns in width or smaller; $PM_{2.5}$ = particulate matter two-and-one-half microns or less in width.

a. There are significant variations within each source category. New plants and stationary and mobile installations fully equipped with scrubbers, filters, and low-NO_x burners have a negligible impact on air pollution. See chapter 5.

b. The column "Consumption goods" includes heterogeneous sources because some goods contribute to chlorofluorocarbon and hydrofluorocarbon emissions (mainly refrigerators), while other goods contribute to volatile organic compound emissions (mainly solvents and paints), although spray aerosols release both.

BOX 4.1

Creative management of trade-offs between large and small sources of air pollution in Krakow, Poland

In the 1980s, the winter smog in the Polish medieval city of Krakow became a major health and social crisis. Initially, public opinion blamed the large industries with visible tall stacks, including two coal-fired combined heat and power (CHP) plants, one steel plant, and one aluminum smelter. The source-apportionment studies and airshed modeling confirmed that one upwind power plant and the aluminum smelter contributed to low air quality in the city, but the steel plant and another CHP plant were downwind and their contribution to urban winter smog was relatively small; so was the contribution of traffic. The public was surprised that the main culprit was low-stack emissions from hundreds of small coal-fired heating boiler houses scattered around the city and thousands of individual coal stoves in peoples' flats and houses. Local environmental authorities, supported by a strong social movement, managed to first close the most polluting industrial processes in the aluminum smelter and then enforce dust pollution control on both CHPs and a sulfur scrubber on the upwind CHP.

However, handling the low-stack pollution sources was difficult and socially sensitive because their environmental retrofit was not economically viable (too small for end-of-pipe controls), while their replacement required mobilization of massive financing for investments in gas, district heating, or electric heating infrastructure and conversion of thousands of individual boilers and stoves. Behavioral change was also needed because individual solid fuel stoves were perceived as being more reliable sources of heat than centrally distributed gas or hot water. Amid the economic downturn after the collapse of the centrally planned economy, public coffers were empty. One of the remaining reliable sources of public revenue for regional and local authorities was pollution fees and noncompliance fines paid by industry and the power sector and the Ecofund, which was created from debt-for-environment swap with a few sovereign lenders

Luckily, the dawn of the market economy came with the phaseout of energy subsidies and was accompanied by the decentralization of authority to regional governors and municipalities. Municipal utilities, including Krakow's district heating company (MPEC) and gas distribution utility, were commercialized and quality of service was increased as a result of competitive pressure to attract customers. It so happened that the downwind CHP faced the deadline to install a flue-gas desulfurization unit but needed more time to attract funds for deep retrofit investments. Missing the deadline meant high noncompliance fines paid from after-tax income. The head of the regional environmental protection department, Jerzy Wertz, saw this as an opportunity to facilitate a deal between the city, MPEC, and the CHP operators under the auspices of a broader Low-Stack Emission Program for Krakow.

This program was supported by Sweden, the United States, and the World Bank, which provided a long-term loan and advisory services to MPEC. Under this deal, the environmental protection department decided not to collect noncompliance fines from the CHP in return for the CHP operator's commitment to transfer an equivalent amount (from pretax income) to the municipal account used by the district heating company to prematurely retire coal boilers and stoves and connect their users to the district heating system at no cost to them. In this way, the CHP increased sales of heat to the city, thereby boosting revenue, which made it easier to secure financing for the flue-gas desulfurization unit a few years later. The World Bank, the Ecofund, and local environmental funds cofinanced the investment program under which MPEC alone scrapped 387 of the most-polluting coal-fired boiler houses in the city center during 1990–2004, significantly improving local air quality and enhancing its own service delivery and commercial viability.

Sources: http://sgpm.krakow.pl/aanewsysn/UserFiles/File/2013-11-15-prez6.pdf; personal experience; and communications with stakeholders.

waste and biomass. These sources would not be the priority for climate mitigation because their impact on global warming is smaller relative to large combustion sources. Climate mitigation would rather begin with large power and industrial sources of emissions, in addition to transport, waste, and agriculture, where opportunities for win-win environmental effects are more common (see also box 4.1).

The elimination and reduction of fossil fuel use offer synergies between improved air quality and slower warming, but not when users switch from gas to biomass, which involves a trade-off. Furthermore, phasing out of fossil fuels is not always the most effective and quickest ways to improve air quality. Indeed, decommissioning of the sources of combustion of fossil fuel, especially coal, would eliminate emissions of both air pollutants and long-lived climate pollutants, as well as short-lived climate pollutants. However, decommissioning, especially of young coal plants, is possible only in the long term and with abundant resources and alternatives. Decarbonization will require an innovation push, subsidies for new low-carbon technologies, massive investments in infrastructure, paying the costs of stranded assets, major behavioral change, and often renegotiation of the basic social contract. Premature deaths from air pollution can often be prevented more quickly and more cheaply by fixing pollution controls on existing combustion installations rather than by replacing them with alternative energy sources. Installing equipment to control air pollution on power plants can effectively minimize emissions of air pollutants, but at the expense of a small increase in carbon dioxide emissions. Therefore, the fact that air pollutants and GHGs are co-emitted from the same sources does not mean that their abatement measures will always generate air pollution and climate synergies. Chapter 5 dives deeper into the choices of abatement measures.

REFERENCES

Agee, M. D., S. E. Atkinson, T. D. Crocker, and J. W. Williams. 2014. "Non-Separable Pollution Control: Implications for a CO_2 Emissions Cap and Trade System." *Resource and Energy Economics* 36 (1): 64–82. https://doi.org/10.1016/j.reseneeco.2013.11.002.

Lvovsky, Kseniya, Gordon Hughes, David Maddison, Bart Ostro, and David Pearce. 2000. "Environmental Costs of Fossil Fuels: A Rapid Assessment Method with Application to Six Cities." Environment Department Paper 78, Pollution Management Series, World Bank, Washington, DC. https://openknowledge.worldbank.org/handle/10986/18303.

Purohit, Pallav, Markus Amann, Gregor Kiesewetter, Peter Rafaj, Vaibhav Chaturvedi, Hem H. Dholakia, Poonam Nagar Koti, et al. 2019. "Mitigation Pathways towards National Ambient Air Quality Standards in India." *Environment International* 133, Part A. 105147. https://doi.org/10.1016/j.envint.2019.105147.

Radu, O. B., M. van den Berg, Z. Klimont, S. Deetman, G. Janssens-Maenhout, M. Muntean, C. Heyes, F. Dentener, and D. P. van Vuuren. 2016. "Exploring Synergies between Climate and Air Quality Policies Using Long-Term Global and Regional Emission Scenarios." *Atmospheric Environment* 140: 577–91. doi:10.1016/j.atmosenv.2016.05.021.

Updyke, Katelyn M., Tran B. Nguyen, and Sergey A. Nizkorodov. 2012. "Formation of Brown Carbon via Reactions of Ammonia with Secondary Organic Aerosols from Biogenic and Anthropogenic Precursors." *Atmospheric Environment* 63 (December 2012): 22–31.

World Bank. 2018. "Local and Regional Pollution Reduction Co-Benefits from Climate Change Mitigation Interventions: A Literature Review." IEG Working Paper 2018-1, Independent Evaluation Group, World Bank, Washington, DC.

5 Abatement Measures

INTRODUCTION

This chapter explores the conditions under which, for a given source, the same abatement measure can achieve a joint—or opposite—impact on air quality and climate. "Abatement measures" are technical or behavioral means that reduce emissions. Abatement measures are distinguished from, and treated as outcomes of, abatement policies, which are the regulatory and financial incentives for implementation of abatement measures and discussed in chapter 6.

Synergies between reducing air pollution and mitigating climate change are more common for the cheapest and the most expensive abatement measures, whereas several medium-cost measures involve trade-offs between air pollution and climate mitigation (see figure 5.1). Several quick and easy win-win opportunities can be found with improved maintenance and small repairs, inspection regimes, tune-up, and better fuel quality in existing sources. Although their impact on air quality and climate co-benefits is usually limited, these measures can be achieved quickly. More expensive end-of-pipe technologies can remove up to 99 percent of pollutants from exhaust fumes of stationary and mobile sources. Their emissions reduction potential is significant (if operated properly), although the capital and operating costs to the polluters are higher and often involve an energy penalty, that is, their operation requires ancillary energy consumption (Poullikkas 2015), thereby increasing carbon dioxide (CO_2) emissions per unit of output. Abatement options such as the retirement of existing combustion sources and replacement with renewable energy sources, or clean technology processes, have the greatest synergy between air pollution and climate benefits because such options eliminate the sources of fossil fuel combustion. However, such measures require high initial capital investment (though operating costs may be lower), long lead times, and often complex support infrastructure (such as sufficiently flexible grid networks to ensure reliable energy supply with variable renewable energy sources).

FIGURE 5.1

Cost-effectiveness of measures to reduce concentration of air pollutants (PM₂.₅)

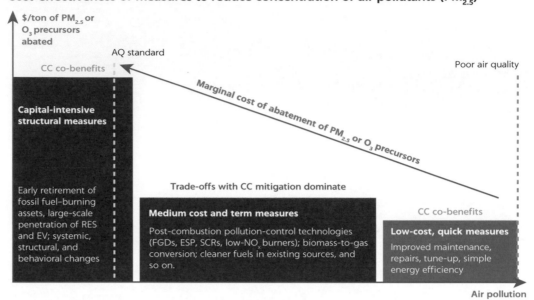

Source: World Bank.
Note: AQ = air quality; CC = climate change; ESP = electrostatic precipitator; EV = electric vehicle; FGDs = flue-gas desulfurization; NO$_x$ = nitrogen oxides; O$_3$ = ozone; PM$_{2.5}$ = particulate matter two-and-one-half microns or less in width; RES = renewable energy sources; SCR = selective catalytic reduction.

Sequencing emission-control measures from the cheapest to the most expensive per unit of reduced pollution exposure is mathematically and graphically represented by marginal abatement cost curves (figures 5.1 and 5.2).[1] The horizontal axis of marginal abatement cost curves represents the reduction of concentration of a pollutant in the air for a specific exposed population. The vertical axis represents the cost of reducing concentration by one unit (usually microgram per cubic meter [μg/m³]). For particulate matter (PM$_{2.5}$) the costs include the cost of reducing emissions of its precursors, such as ammonia, nitrogen oxides (NO$_x$), volatile organic compounds (VOCs), or sulfur dioxide (SO$_2$). Therefore, in a given location, each abatement measure—for example, banning diesel cars in the urban center or installing end-of-pipe equipment—is represented by a rectangular shape (figures 5.1 and 5.2), where length is its potential to improve quality and height is the marginal cost of realizing this potential. Some of these measures will have climate co-benefits, while others will increase emissions of greenhouse gases (GHGs) and, hence, present trade-offs for abatement of air pollutants and GHGs.

Managing multiple environmental crises is particularly challenging in developing countries and cities grappling with limited resources and institutional capacity. The proposition of implementing emission-abatement measures that simultaneously address both air pollution and climate change sounds attractive on its face, but only if addressing both problems does not overwhelm local budgets and capabilities (Dulal and Akbar 2013). Policy makers in developing countries are careful about packaging and sequencing abatement measures to save the maximum number of lives with available resources. Prioritizing expensive air quality improvement measures, such as

FIGURE 5.2

Prioritizing air quality improvement with and without climate co-benefits

a. Air quality–led package of abatement measures

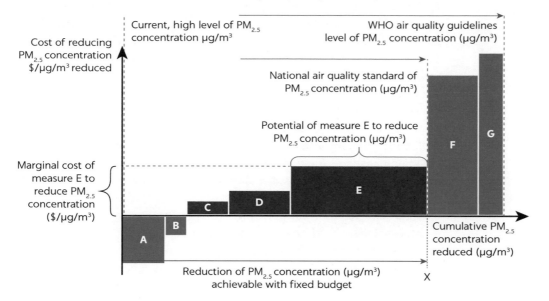

b. Climate co-benefits–led package of abatement measures

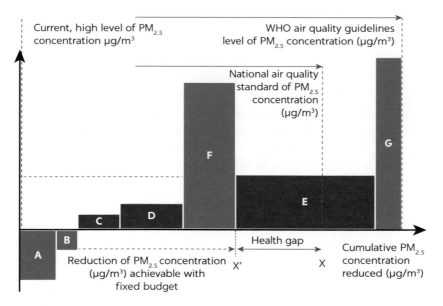

Source: World Bank.

Note: Orange bars represent those measures to reduce particulate matter ($PM_{2.5}$) concentration in the target airshed that show climate co-benefits, such as energy efficiency (bars A and B) and structural changes involving premature retirement of fossil fuel assets (bars F and G). Red bars are $PM_{2.5}$ concentration-reduction measures without climate co-benefits, such as switching from biomass to natural gas or end-of-pipe filters. X is a least-cost environmental effect (reduction of population exposure to $PM_{2.5}$, achieved with a given, fixed resource constraint. X′ is the lower environmental effect that can be achieved with the same budget constrained by using measure F with climate co-benefits instead of measure E. WHO = World Health Organization; $\mu g/m^3$ = micrograms per cubic meter.

massive, accelerated retirement of fossil fuel combustion sources and replacement with renewable energy, can enhance climate co-benefits but may also increase the total cost of improving air quality. These higher costs could lead to the exhaustion of available resources before safe air quality standards are attained. This prospect is illustrated by panels a and b of figure 5.2. Therefore, an integrated approach to policies on air quality and policies on climate change mitigation requires more than just prioritizing win-win measures. A well-intentioned focus on win-win abatement measures can reduce or delay the health benefits of an air quality program, especially under a hard budget constraint. It should also be said that sometimes developing countries embark on the excessively expensive measures for reducing health impacts from air pollution for other reasons. For example, sometimes they apply high European emissions standards for vehicles, even though the abatement measures with climate co-benefits (such as modal switch to public transit or bicycles) are cheaper.

Figure 5.2 should be read as follows: The step function represents the marginal cost of reducing the concentration of air pollutants ($PM_{2.5}$) in a targeted airshed. The length of each step (rectangle) represents the potential of this measure to reduce PM concentration. The height represents the average marginal cost (after any revenue or savings) per unit of improved air quality. The rectangles located in negative territory (below the x-axis) have negative costs, meaning that they yield positive net economic returns even before the avoided costs of pollution damage are considered. Abatement measures that look to have negative costs (often associated with energy efficiency) still require upfront investments and often face hidden costs and barriers not visible to traditional marginal abatement cost curve models but that are clearly visible on the ground (Vasquez Suarez, Liu, and Peszko 2018). The area of each rectangle shows the total cost (or total net economic return if negative) of implementing the measure. Total abatement cost is represented by the combined areas of all the red and orange rectangles. Orange rectangles represent measures to abate air pollution that have climate co-benefits. Red rectangles represent measures that reduce $PM_{2.5}$ pollution without climate co-benefits, or with some warming effect (for example, by switching from biomass to gas in household cooking and heating or installing end-of-pipe emission controls).

In panel a of figure 5.2, the least-cost program to attain national air quality $PM_{2.5}$ standards consists of measures A, B, C, D, and E, which jointly improve air quality to the national standard X. The keys to achieving the least-cost improvements to air quality are measures C, D, and E, which show negligible climate co-benefits or even slightly increase carbon emissions. All these measures are within the budget available for the air pollution program. However, should the policy makers decide to implement measure F, which has a large climate co-benefit (for example, switching to electric cooking and heating), they would have to give up measure E, because both are not affordable given the hard budget constraint. Measure F has a similar total cost, but can reduce less pollution than measure E, because of a higher cost per unit of abatement of $PM_{2.5}$ pollution. Consequently, much less air quality improvement is affordable—only X′ in panel b of figure 5.2, leaving a health gap of X – X′. International development partners may provide partial financial assistance to leverage implementation of measures F and G because they provide not only local but also global public good (climate co-benefits). These measures can also be implemented at the extra domestic cost in the name of some other self-interest for the host country, such as higher demand for air quality, modernization of the economy, jobs creation, or some other benefits from participating in the cooperative climate action.

MAPPING SYNERGIES AND TRADE-OFFS BETWEEN ABATEMENT MEASURES FOR AIR POLLUTION AND CLIMATE CHANGE

Energy efficiency plays an important role in the early stages of air quality programs, followed by switching to cleaner fuels, but major improvements require end-of-pipe abatement technologies. Rafaj, Amann, and Siri (2014) apply decomposition analysis and find that three-quarters of the decline in SO_2 emissions in Western Europe during 1960–2010 resulted from a combination of reduced energy intensity, structural changes, and improved fuel mix, while end-of-pipe abatement measures were responsible for 22 percent of the decoupling of SO_2 emissions from economic growth. End-of-pipe measures played a dominant role in the reduction of NO_x emissions, however. Looking at more recent trends, between 2000 and 2010 Rafaj et al. (2014) show that in Western Europe, end-of-pipe emission control was responsible for as much as 50 percent of the decoupling of emissions of all air pollutants from economic growth, while 35 percent was driven by changes in fuel mix and 15 percent by energy intensity and energy efficiency improvements. Decomposition analysis in EEA (2019) for the European Union (EU) confirms that between 2004 and 2015, end-of-pipe controls and the switch to low-sulfur and low-dust coal accounted for 75 percent of dust, 71 percent of SO_2, and 38 percent of NO_x emissions reductions from large electricity-generating combustion plants. None of these measures show climate co-benefits. Reduction of energy intensity of the economy and the switch away from coal played relatively small roles (though higher for NO_x; 9 percent and 15 percent, respectively), while these two drivers dominated reduction of CO_2 emissions.

Andaloussi (2018), in an empirical statistical decomposition of the factors driving reductions in emissions of local air pollutants in the US power sector between 2005 and 2014 at the plant level, finds that the adoption of capital-intensive end-of-pipe abatement technologies accounted for more than 50 percent of the achieved reductions. In addition, Massetti et al. (2017) find that from 1994 to 2004, the majority of reduction of SO_2 emissions from the US electricity sector resulted from the application of flue-gas desulfurization technology. Switching to cleaner fuel inputs and retiring dirty units each contributed about 20 percent of the observed reductions. Energy efficiency was not included in this decomposition analysis. Fuji, Managi, and Kaneko (2013) find that end-of-pipe measures were the dominant driver of changes in SO_2, dust, and soot emissions from the Chinese industrial sector from 1998 to 2009, although energy efficiency also played an important role for SO_2 and soot emission trends, while dust emissions were reduced mainly by end-of-pipe measures and improvements in production processes. Iyu et al. (2016) also find that in China, energy intensity was a stronger driver of the trends of emissions of particulates and their precursors than were changes in the fuel mix. Rafaj and Amann (2018) observe that in China, India, and Japan, fuel switching remains difficult and that, historically, energy intensity was the key driver of changes in air emissions. However, as countries began using energy more efficiently, the potential for energy efficiency measures to prevent air pollution began decreasing, and most of the remaining potential depends on end-of-pipe control measures, the phasing out of the combustion of fossil fuels, or both.

The local and temporary trade-offs between air pollution reduction and climate change mitigation are as common as win-win abatement measures available to economic agents. Figure 5.3 illustrates four common pollution abatement measures that usually deliver climate co-benefits (in the top-right quadrant) and three that have ambiguous or a slightly warming impact on climate (top-left quadrant). Two climate-mitigation measures in the bottom-right quadrant should be watched for their potential to deteriorate air quality. The presence of occasional tensions is a fact that must be acknowledged and managed. It is not an excuse to delay action on either the local health threat of air pollution or the existential threat of climate change.

Researchers from the Global Alliance on Health and Pollution (GAHP), AirQualityAsia, and the Schiller Institute for Integrated Science and Society at Boston College (2020) evaluated 22 interventions intended to reduce air pollution. Using their expert judgment, they found those interventions that both improve local health and favorably affect climate change (such as replacing fossil fuels with renewable energy sources) are of limited value for both environmental problems. However, the interventions that would increase CO_2 emissions were not included in the assessment (such as the postcombustion air pollution abatement controls), or were not quantified (such as the climate-warming impact of reducing SO_2 and NO_x emissions).

FIGURE 5.3

Synergies and potential tensions between key mitigation measures for air pollution and climate change

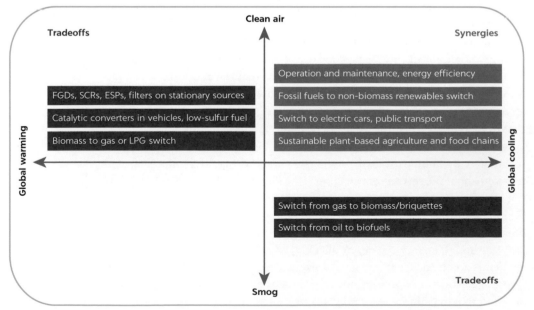

Source: World Bank.
Note: ESPs = electrostatic precipitators; FGD = flue-gas desulfurization; LPG = liquefied petroleum gas; SCR = selective catalytic reduction.

After correcting for these missing interventions, the trade-offs in that study would be more aligned with figure 5.3.

Win-win measures for air quality and climate mitigation

Improved operational efficiency and maintenance of vehicles, stationary combustion installations, or entire infrastructure systems can improve both their fuel efficiency (carbon emissions) and local environmental performance at relatively low incremental cost. Therefore, these measures should be an essential component of any integrated air pollution and climate change program. They do not require significant upfront investments, although maintenance expenditures for equipment operators may not be trivial. Hence, enforcement can be challenging, especially in lower-income communities with weak institutions, even if maintenance costs are often more than offset by future lower operating costs. However, their potential to achieve both climate and air quality objectives is limited so additional abatement measures are necessary. In addition, improved management of transport systems, through regulatory restrictions, low emission zones, parking controls, delivery planning and freight route optimization, and improved traffic flow management, improves local air quality and also has a climate-mitigation impact (Khreis, May, and Nieuwenhuijsen 2017; Letnik et al. 2018; Slovic et al. 2016; Xia et al. 2015).

Switching from fossil fuels to noncombustible renewable sources of energy has a double environmental dividend because it reduces combustion activities and all global and local emissions related to them. However, it is often a costly abatement measure and lengthy to implement, especially if existing assets based on fossil fuels are relatively new and have many years of economic life left. Large-scale and quick phasing out of fossil fuel-fired emission sources may not be feasible in all local circumstances if renewable resources are not locally available at scale, not cost-competitive, or not yet enabled by required infrastructure, such as power lines, flexible resources in the grid to manage variable renewable energy, or a charging network in transport. In some industrial sectors, such as steel or cement, the phasing out of fossil fuel combustion is even more expensive and time consuming. Therefore, these measures may not always be the first choice of policy makers in developing countries who are under pressure to achieve rapid improvement of air quality at low cost to prevent premature deaths and disease from year to year.

Energy and resource efficiency deliver a double environmental dividend. Burning less fuel to deliver the same level of service or output prevents local pollution and GHG emissions. In many marginal abatement cost models, energy efficiency is portrayed as a negative-cost measure. Although it has clearly been the case in the countries with massive legacies of energy-wasteful, inefficient infrastructure (for example, in the countries of the former Soviet Union or rapidly industrializing China and Asia in the 1990s), in many countries the remaining energy-efficiency options are more expensive and difficult to implement because of high transaction costs, relatively low returns aggravated sometimes by low regulated prices, and difficult access to finance; all of these factors are often neglected in the models that show costs and benefits from the point of view of an abstract social planner, rather than specific economic actors. A comprehensive package of policies, including those that correct energy prices to cover external environmental costs are needed to make socially optimal technical measures profitable to economic actors. Such policies are also needed to prevent rebound effects from offsetting significant climate benefits of energy efficiency measures (Vasquez Suarez, Liu, and Peszko 2018).

Other technical and behavioral measures that typically have double benefits for climate and air pollution include:

- *Modal switch in transport.* Shifting passenger and freight transport from individual vehicles to public transport and from roads to rails usually delivers both climate and air quality benefits. The scale of the benefits depends on the emissions intensity of both individual and public vehicles and the ability to induce behavioral change in firms and commuters (also discussed below).
- *Nature-based solutions* harness ecosystems' capabilities to provide the same services as the built infrastructure. Therefore, they always yield multiple air quality and climate benefits. However, they do not always displace the key sources of anthropogenic air and climate pollutants.
- *Consumers' behavioral changes.* Consuming fewer disposable items, changing commuting habits, switching to a mainly plant-based diet, and increasing active travel (walking and cycling) usually have multiple benefits, including benefits regarding air pollution and climate, as well as for personal wellness and health (Woodcock et al. 2009; Xia et al. 2015).
- *Banning open burning of biomass and waste.* This is one of the major sources of particulate matter (PM) and carcinogenic emissions in many developing countries. It also releases some CO_2 to the atmosphere, but the climate co-benefits of reducing it are small, if any, and depend highly on local circumstances because the carbon emitted through burning would otherwise be partially trapped in the soil and partially released as methane, which is a potent GHG. Therefore, preventing the burning of municipal and agricultural waste, and preventing the disposal of plastic, paper, and organic waste, have much higher benefits to air pollution than to the global climate.

Measures to abate air pollution that warm the climate

Switching from biomass to natural gas in household heating and cooking saves lives from air pollution in developing countries with a small, warming impact on the climate. Worldwide, between 2.5 billion and 3 billion people or more still rely on traditional biomass for cooking and heating. Household air pollution from cooking alone kills between 3 million and 4 million people every year, more than malaria and tuberculosis combined (ESMAP 2020; WHO 2014). Switching from biomass to gas is a quick and effective way to reduce particulate emissions from small boilers and cookstoves, with significant benefits to health, especially for indoor applications that pose major health hazards, including to women, children, and the elderly who spend more time at home. However, switching from biomass to liquefied petroleum gas (LPG) or natural gas has been controversial because renewable biomass is replaced by fossil fuels. In line with international and European Union (EU) common practice, CO_2 released from the combustion of biomass is accounted for as zero in carbon footprinting methodologies as long as this biomass comes from sustainably managed forests (Regulation (EU) 2018/841, Directive (EU) 2018/2001, Regulation (EU) No 601/2012, and Regulation (EU) No 525/2013). This cannot be said about natural or petroleum gas, whose carbon footprint is about half that of coal per unit of useful energy. It is recognized that biomass combustion may release other GHGs, such as methane and nitrous oxide (N_2O) as well as black carbon (BC), organic carbon (OC), and carbon monoxide (CO), especially from small inefficient stoves and boilers (EIB 2020). In developing countries, biomass used in small household stoves and small boilers often does not come from sustainable sources, hence is sometimes considered nonrenewable, although most studies find that fuel wood is not the major driver of deforestation (see discussion in Lee et al. 2013). A literature survey conducted by Pittel and Rübbelke (2008) concludes that displacing biomass by natural or petroleum gas in combustion sources increases GHG emissions while decreasing emissions of NO_x and PM. More recent life-cycle assessments of the climate impacts of switching from biomass to gas in household cooking and heating agree that the cumulative warming impact on climate is negligible when gas is compared with currently used stoves burning unsustainable biomass (Bruce, Aunan, and Rehfuess 2017; IEA 2017; Kypridemos et al. 2020). In addition, Pachauri, Rao, and Cameron (2018) show that, because of the efficiency gains of switching from biomass to fossil fuels for cooking, the associated climate penalty is very small. The short-term temporary effect on climate was even found to be positive in some locations and where biomass burning led to significant emissions of BC as opposed to OC (Rosenthal et al. 2018; Singh, Pachauri, and Zerriffi 2017).

Climate and carbon finance are often not available for biomass-to-gas conversion even for household cooking and heating. Neither voluntary (Gold Standard) nor United Nations Framework Convention on Climate Change Clean Development Mechanism methodologies exist for displacing biomass by LPG or liquefied natural gas (LNG). Even conventional development assistance sometimes is not available for such projects if climate co-benefits are among the eligibility criteria. Interestingly, among 57 carbon offset projects included in the Catalog of Carbon Offset Projects of the Global Alliance for Clean Cookstoves, two involve LPG—one was switching from biomass to LPG and the other from LPG to biomass briquettes, both achieving very small GHG emissions reductions compared with other projects that usually improved the efficiency of biomass stoves or displaced biomass with other renewable sources (GACC 2014). Some providers of climate finance (for example, the multilateral development banks' technical working group on climate finance methodologies) are now considering household cookstove biomass-to-gas conversion projects if substantial net GHG emissions reductions are demonstrated at a project level. The requirement for project-by-project proofs limits the scale and increases transaction costs of obtaining climate finance for switching from biomass to gas for household cooking and heating in developing countries.

End-of-pipe pollution-control measures can effectively improve local air quality but often at the cost of some increase in climate warming. The trade-off between the impacts on air pollution and climate has two channels. The first channel is through the reduction of emissions of climate coolants. End-of-pipe measures to abate air pollution often reduce emissions of precursors of sulfate and nitrate aerosols that, as discussed in the previous chapter, have cooling effects on climate. A second channel is through auxiliary energy consumption by equipment to control air pollution. Adding devices such as particulate filters, SO_2 scrubbers, or catalytic reduction of NO_x restricts stack flow and requires increases in plants' on-site consumption of electricity (the so-called energy penalty), resulting in an increase in CO_2 emissions per kilowatt hour (kWh) of final useful energy generated (EPRI 2011). The energy penalty may amount to up to 5 percent of gross power output when ancillary electricity consumption of systems for all air pollutants are accounted for (Burnard et al. 2014; Cropper et al. 2017; EPRI 2011; Masters 2004; Rubin and Nguyen 1978; Srivastava and Jozewicz 2001; Srivastava et al. 2005).

The World Bank assessment of green growth opportunities for North Macedonia in 2014 finds that significant PM emissions from industrial sources in North Macedonia can be reduced without climate co-benefits. More than 90 percent of particulate emissions are generated by five economic activities, and 92 percent come from the largest industrial facilities. "Particulate matter pollution could be reduced by up to 80 percent in the ferroalloys and energy sectors and by 90 percent in road paving activities and among wood-burning households by using simple technologies: adoption of dust collection and scrubbing technologies in the energy sector; usage of dust collectors in road pavement processes; and replacement of inefficient wood stoves, increasingly used for heating in response to tariff increases, in the household sector.... For example, in the biggest polluter Jugohrom Ferroalloy's, emissions could be reduced by up to 80 percent through such measures as low energy scrubbers, sealed furnaces and enclosed product transfer (for example, conveyor) systems. Emissions from public electricity and heat production could be reduced up to 80 percent through the installation of pollution abatement equipment and fuel switching (from lignite to gas)" (World Bank 2014, 169–75). Only the last measure (switching from coal to gas) has climate co-benefits. Other measures would effectively reduce particulate emissions with some warming or neutral impact on climate.

The most common end-of-pipe pollution-control installations with a climate-warming effect include the following:

- Flue-gas desulfurization units, so-called scrubbers, capture 90–99 percent of sulfur emissions from exhaust fumes, depending on the technology used. Scrubbers reduce emissions of a precursor of sulfate aerosols, which are climate coolants, as discussed in chapter 3. They also add an energy penalty, which can vary between 0.7 percent and 2 percent of gross power generation, thus increasing CO_2 emissions intensity of a coal power plant accordingly (Forbes 2018; Poullikkas 2015; Sargent & Lundy 2009).

- Filters for PM include fabric filters and highly effective electrostatic precipitators that can reduce more than 99 percent of PM emissions from stationary coal-combustion sources (see figure 5.4), including fine particles at state-of-the-art installations. Particulate filters on stationary sources may have an energy penalty of greater than 1 percent. Since 1985 diesel engines have come equipped with diesel oxidation catalysts and diesel particulate filters. Diesel oxidation catalysts use NO_2 to oxidize particulate emissions, increasing NO_x emissions (ICCT 2016). Diesel particulate filters are highly efficient in removal of PM from exhaust fumes, but backpressure reduces engine efficiency, and additional costs of filter regeneration and maintenance are involved.[2] Therefore, they are sometimes (illegally or not) removed by vehicle owners in developing countries, emerging economies, and even in the European Union (Carrington 2016).

- NO_x abatement technologies, such as selective catalytic reduction, reduce emissions of a precursor of nitrate aerosols—which are climate coolants and slightly increase emissions of CO_2—through auxiliary internal energy consumption (Burnard et al. 2014). Other measures, such as fluidized bed boilers, by lowering the temperature in combustion chambers, give rise to

FIGURE 5.4

Emissions of major air pollutants from new coal power plants having, versus those not having, state-of-the-art emission-reduction installations

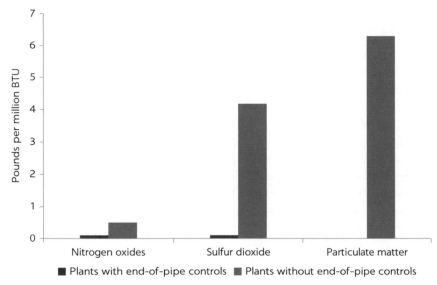

Source: World Bank, based on IER (2017) (data from National Energy Technology Laboratory.
Note: BTU = British thermal unit.

increased emissions of other pollutants, such as N_2O, which is a GHG (Richards 2000). Selective catalytic reduction in vehicle engines slightly decreases efficiency of the engine, marginally increasing fuel use and CO_2 emissions per kilometer traveled, hence sometimes also leading to their (usually illegal) removal (although the primary motivation is recovery of precious metals). Moreover, catalytic converters (especially older ones) in vehicles produce N_2O as a result of incomplete reduction of NO_x to nitric oxide (ARB 2014). However, integrated simultaneous regulation of air and climate pollutants (such as in California and the European Union) encourages the vehicle technology that turns the trade-offs between NO_x and GHG emissions from internal combustion sources into synergies. For example, the US-based Manufacturers of Emission Controls Association reports that a review of heavy-duty engine certifications from 2002 to 2015 shows that once emission and efficiency technologies were required on engines, the relationship between CO_2 and NO_x emissions went from a trade-off to a mutual benefit. By setting stringent emissions targets for both CO_2 and NO_x through regulations and expansion of the calibrator's toolbox from the engine to the powertrain allowed engineers to achieve reduced NO_x levels and engine efficiency improvements simultaneously, thereby boosting technology innovation and cost savings.[3] Agee et al. (2014) make a similar observation for US power utilities.

End-of-pipe installations can be expensive. In Organisation for Economic Co-operation and Development countries, up to 40 percent of the cost of building a new coal power plant is spent on equipment to control emissions (EPRI 2011; Masters 2004). Therefore, such investments are usually either integrated into the design of new plants or incorporated into the complex retrofitting of older plants. Additional capital and operational cost require long-term returns and high annual operating hours to break even. Deep retrofits increase the asset value of the entire thermal energy installation, making its early retirement for climate reasons more economically and socially challenging. The operating and maintenance costs of emission-control equipment are also high. Particulate control can add 2 percent to the cost of electricity, while flue-gas desulfurization and selective catalytic reduction may increase the cost of electricity by about 10 percent and 6 percent, respectively (Wu 2001). End-of-pipe installations also show strong economies of scale and are usually applied in larger power, heating, and industrial plants (IER 2017). This is important for an integrated policy approach because, in the absence of climate policy incentives, significant amounts of capital can be locked into plants that contribute little to air pollution, but that can be stranded prematurely because of the impacts of the low-carbon transition (Peszko, van der Mensbrugghe, and Golub 2020; Spencer, Berghmans, and Sartor 2017). The opposite approach is also risky: in the absence of effective air pollution regulations, the pressure to minimize the climate impact sometimes leads to investments in very efficient and expensive fossil fuel installations (for example, ultra-supercritical coal–fired power plants), but with substandard air pollution–control equipment. Furthermore, higher carbon and energy prices put economic pressure on companies to reduce the use of air pollution control technologies even if they are installed. The economic analysis of new fossil fuel investments (whether new or retrofits) should always consider the costs of both state-of-the-art fuel and carbon efficiency and state-of-the-art pollution controls.

Switching from diesel to gasoline engines improves air quality but without significant climate benefits. Diesel fuels have a slightly higher carbon content per unit of energy, but diesel engines tend to be more efficient, so the GHG emissions per kilometer were for a long time lower than in gasoline engines (Edwards et al. 2014). However, emissions of particulates and carcinogenic aromatic hydrocarbons, including BC, are higher. Historically, diesel engines have been promoted (for example, through lower excise and vehicle taxes) as the more efficient and low-carbon alternative to gasoline engines (Government of India 2010; OECD 2019). For example, vehicle tax reform in Ireland introduced in 2008 successfully improved the fuel economy of new passenger cars but also increased the proliferation of diesel vehicles in the passenger car fleet and increased emissions of local air pollutants (Ryani et al. 2019). Inconsistent priorities for air pollution and climate change in transport have led to some confusing and expensive policy reversals in many European countries, including the United Kingdom. In the 2000s, the UK government had a policy of increasing the real level of taxes on gasoline and gasoline-powered vehicles more heavily than diesel vehicles. This strategy seemed to offer a convenient way to increase tax revenues and reduce CO_2 emissions from the vehicle fleet. Higher fuel taxes and preferences for diesel led to exceedances of local air quality standards for NO_x and particulates. The British government felt that politically it had to halt and then reverse the preferences for diesel in real fuel taxes, while local authorities started to penalize the ownership and operation of diesel vehicles, especially private cars, to meet tightening air quality standards.[4]

Similarly, switching from fuel oil to biofuels may cut CO_2 emissions but increase air pollution. Biofuels, especially second generation (sustainably sourced) ones, are counted as having zero carbon emissions but often imply an increase in NO_x, PM, CO, and VOC emissions (Brännlund and Kriström 2001). A similar observation is made by Portugal-Pereira et al. (2018) for switching from coal to biomass in Brazil's power plants, discussed in the section Modeling Interlinkages between Air Pollution and Climate-Mitigation Measures.

The warming potential of some air pollution abatement measures is by no means an excuse for not pursuing them. Such measures can be very effective at protecting human health and mitigating the negative impact of pollution on ecosystems. Modern coal power plants with state-of-the-art equipment to control air pollution have a negligible impact on exposure to SO_2, NO_X, and PM (see figure 5.4). In mobile sources, filters and catalytic converters also minimize the impact on ground-level ozone formation and $PM_{2.5}$ exposure. Therefore, end-of-the-pipe installations to reduce emissions of air pollutants from fuel combustion installations and engines are essential elements of modern economy. Such installations are embedded in environmental management regulations around the world, including the best available techniques (as defined in the EU Industrial Emissions Directive) and best available technologies or similar environmental performance standards defined in other jurisdictions (OECD 2020). An integrated approach to air pollution and climate change requires that the inevitable trade-offs for some abatement measures be acknowledged and managed by parallel and coordinated regulation of air pollutants and GHGs emitted by the same sources, with each measure based on its own merits. In this way, investors and operators can make informed decisions about planning new investments or retrofitting existing sources—whether to invest in low-air pollution fossil fuel sources or to choose noncombustion technologies instead. Such integrated investment appraisals can accelerate the retirement of fossil fuel combustion installations or boost technology innovation for synergistic solutions.

Relocation of local pollution sources can alleviate local air pollution but does not mitigate climate change. Moving sources of air pollution downwind of population centers or increasing the height of stacks to disperse air pollutants can be effective measures for reducing population exposure. In principle, these are not good air quality management practices, but sometimes can be used creatively to bring about quick temporary relief from smog (see box 6.3). Relocating emission sources (horizontally or vertically) makes no difference to climate change caused by uniformly dispersed GHGs (such as CO_2, hydrofluorocarbons, N_2O, or methane), though doing so may relieve short-term warming by relocating BC emissions to surfaces where it has lower climate forcing. In addition, these measures can shift air pollution from one place to another—for instance, decreasing the exposure of the local population while increasing the exposure of distant fragile ecosystems to eutrophication and acid rain. In coal-dependent countries, electric mobility moves pollution from low-exhaust mobile sources to large stationary coal power plants with tall stacks and sometimes downwind locations. This benefits air pollution mitigation, but under some conditions, contributes to climate warming. Box 5.1 discusses the dilemma of prioritizing several available measures to reduce air pollution related to vehicle emissions in Tehran.

Measures to mitigate climate change that can increase air pollution

Many climate-mitigation pathways compiled with integrated assessment models envisage a major increase in the role of bioenergy in the world's energy balance (Huppmann et al. 2019; Rogelj, Pop, et al. 2018; Rogelj, Shindell et al. 2018; Rogelj et al. 2019). Although the volumes of bioenergy used globally will

From source apportionment to identification of measures to improve air quality in Tehran

Source apportionment of fine particulate matter (PM) pollution in Tehran shows that mobile sources account for between 50 percent (Taghvaee et al. 2018) and 70 percent (Shabhazi et al. 2016) of PM concentrations in the ambient air (see figure B5.1.1). Heavy-duty vehicles (HDVs) account for nearly 85 percent of PM emissions from the vehicle fleet (Heger and Sarraf 2018; Shabhazi et al. 2016).

Thus far, Tehran has taken several measures to improve air quality. These measures include metro expansion, a bus rapid transit system, a low emissions zone with fines for vehicles that have not been maintained and inspected, reduction of the use of high-sulfur fuels, the switching of some vehicles to natural gas, diesel-particulate filters for all new vehicles in 2016, and financial incentives for hybrid and electric vehicles. The World Bank, together with the government, identified the following priority measures to further improve air quality, ranked by their financial costs, effectiveness at improving air quality, and time required for their implementation (Heger and Sarraf 2018):

1. *HDV replacement, including a scrappage component*
2. Diesel particulate filter retrofits for HDVs
3. Expansion of the low emissions zone
4. Improved inspection and maintenance system
5. *Provision of incentives for electric and hybrid vehicles*
6. *Provision of incentives for nonmotorized transport*
7. *Expansion of bus rapid transit lines*
8. *Expansion of metro lines*
9. *Strengthened monitoring, measurement, and analytical capacity*

Measures in italics are likely to benefit both air pollution reduction and the climate. Interestingly, out of the top five priority measures, four are expected to deliver low-cost and quick improvements of PM concentrations, but negligible climate co-benefits. The first measure, accelerated replacement of HDVs, may or may not demonstrate climate co-benefits, depending on how much black carbon the old diesel engines were emitting and the vehicles that will replace them.

FIGURE B5.1.1

Sources of PM pollution in Tehran

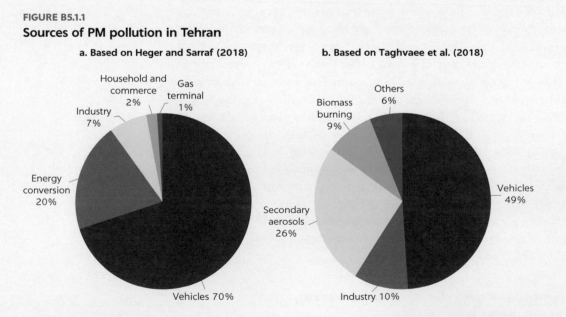

a. Based on Heger and Sarraf (2018)

- Household and commerce 2%
- Gas terminal 1%
- Industry 7%
- Energy conversion 20%
- Vehicles 70%

b. Based on Taghvaee et al. (2018)

- Others 6%
- Biomass burning 9%
- Vehicles 49%
- Secondary aerosols 26%
- Industry 10%

Sources: Based on Heger and Sarraf (2018); Shabhazi et al. (2016); and Taghavee et al. (2018) (underlying data are averages of results from two locations).

continued

Box 5.1, *continued*

The climate benefits of switching to electric and hybrid vehicles depend critically on how electricity is produced, and the emissions related to the transport of fuel to the power plants and tanks. Edwards et al. (2014), in their comprehensive study for the European Commission, find that the European Union's switch to battery-powered electric vehicles would always lead to reductions in greenhouse gas emissions compared with conventional gasoline internal combustion engine vehicles, except in the case in which coal electricity is used. Interestingly, even if all electricity were produced by coal-fired power plants, the lifecycle (well-to-wheel) greenhouse gas emissions of

electric cars in the European Union would be comparable to that of gasoline cars because the coal in the European Union is local, while oil is assumed to be transported for 4,000 kilometers, with a high associated carbon footprint. In the Islamic Republic of Iran, oil would be local, so the carbon footprint of gasoline internal combustion engines would be lower than in the European Union, but if electricity is produced with heavy fuel oil, its carbon footprint would also be high. Measures 6 through 8 relate to a modal switch and will have the strongest climate (as well as accident and congestion) co-benefits but require massive investments and long lead times.

Source: Based on contribution by Martin Heger.

be much lower than the volumes of fossil fuels used now, it is a major source of health risk in the most polluted regions that rely on solid biomass for heating and cooking. Biofuels used in transport can also increase $PM_{2.5}$ exposure in urban areas or in the vicinity of ports, airports, and other major transport hubs. Some climate cooling geoengineering options proposed to scatter the sunlight reaching the earth include atmospheric injection of sulfate and nitrate aerosols, which could lead to additional direct air pollution, indirect pollution through secondary formation of $PM_{2.5}$, and damage due to acid rain and eutrophication of inland waters.

MODELING INTERLINKAGES BETWEEN AIR POLLUTION AND CLIMATE-MITIGATION MEASURES

The argument that air pollution reduction is a co-benefit of climate-mitigation measures

Most climate-led studies focus on calculating the reduction of air pollution as a co-benefit of measures to mitigate climate change. The underlying logic built into such models is that reduction of GHG emissions is proportional to reductions in the use of fossil fuels, thus also limiting the emission of air pollutants associated with fuel use. This would make climate policy attractive from the local perspective (Nam et al. 2014). The models are designed to optimize the cost of GHG mitigation and calculate air quality improvement as a by-product of the expected lower fuel use. Emissions of air pollutants are represented by emission factors linked to fuel use, so any policy that suppresses fuel use by design automatically reduces local emissions. Some models are further linked to simplified air pollution dispersion models, such as TM5-FASST (simplified because they do not consider local topography and meteorological conditions) to estimate changes in concentrations of air pollutants in the air and the resulting associated avoided premature deaths and diseases. The models sometimes also monetize these avoided health impacts and add them to the benefits of reducing GHGs.

Most climate-led models misrepresent the impact of climate mitigation on air pollution. Typically, at least three essential interactions or impact channels between these two types of abatement measures are beyond the scope of the models. First, existing models do not consider explicitly postcombustion air pollution abatement technologies as alternatives to phasing out fossil fuels. The end-of-pipe air pollution control equipment can be an effective and cost-efficient way of reducing emissions of air pollutants from large combustion sources, but it slightly increases their CO_2 emissions. Second, the climate-led models rarely calculate the climate warming caused by reducing sulfate and nitrate aerosols, which is significant, as discussed in chapter 3. Third, most of these models also do not consider air quality impact of switching from natural gas or fossil fuel-fired electricity to renewable energy, such as biomass. Such a switch mitigates climate change but increases exposure of the most vulnerable population to air pollution.

Running models in this limited way, Markandya et al. (2009) and Markandya et al. (2018) formulated a popular climate advocacy proposition that global climate mitigation is in the local national interest because of the air pollution co-benefits it delivers. In economic terms this proposition argues that the economic value of local health co-benefits outweighs the policy cost of achieving the global climate targets. In a similar manner, West et al. (2013) estimate the global average marginal co-benefits of avoided air pollution–related mortality to be US$50–US$380 per ton of CO_2 and conclude that it exceeds the marginal CO_2 abatement costs until 2100. Nemet, Holloway, and Meier (2010) review 37 peer-reviewed national studies conducted with the same approach and estimate that the mean and median values of such collateral air pollution–related health co-benefits of measures are in the range of $44 per ton of CO_2 (tCO_2) and $31/$tCO_2$, respectively, for developed countries, and $81/$tCO_2$ and $43/$tCO_2$, respectively, for developing countries. More recently, such a limited modeling approach has been adopted by Li et al. (2018), Rauner et al. (2020), and the International Monetary Fund (IMF) and part of the World Bank (Coady et al. 2017; Coady et al. 2019; IMF 2019; Parry et al. 2014); Scovronick et al. 2019 and World Bank 2022 apply standard climate-driven modeling approach but explicitly incorporate the cooling impacts of sulfate aerosols into the modeling framework. Scovronick et al. (2019) are transparent that the costs and benefits of targeted air pollution policies as well as their interaction with climate policies are beyond the scope of the modeling. They also make an important caveat that global health benefits from climate policy will depend heavily on the air quality policies that nations adopt independently of climate change, and the authors do not integrate the interactions between climate mitigation and air pollution abatement measures into the model. More recently, Hamilton et al. (2021) use the standard climate-led scenario approach to calculate health co-benefits of implementing countries' nationally determined contributions (NDCs) by 2040. As expected by model design, they find major long-term air pollution health co-benefits from eliminating fossil fuels. This study (like World Bank 2022) assumes that more-ambitious targets for air quality and health policies are achieved independently of the climate policies.

The argument that climate change mitigation is a co-benefit of air pollution measures

Fewer empirical and modeling studies explore climate mitigation as a side effect of air pollution abatement. Those that do often find that air pollution control measures accelerate temperature increase in the short to medium term because of the three interactions between air pollution control and climate mitigation measures discussed at the beginning of this modeling section: climate penalty and cooling effect of end-of-pipe air pollution control technologies and switching from gas to biomass. Originally, there was no need to advocate for air pollution policies in the name of their co-benefits for climate. Local health benefits were making the case strong enough to mobilize action. However, over time, as the climate crisis became better understood, international development institutions (bilateral donors and multilateral development banks in particular) started making international development assistance conditional on the benefits to climate change. Therefore, developing countries wanting to access funds for air pollution mitigation often had to demonstrate that climate mitigation benefits would also be delivered. Modelers rushed to help them. Simulations led by air pollution priorities used the same cost-saving co-benefit argument as climate-led models, but with reverse causality.

Bollen and Brink (2012) conclude (using the computable general equilibrium model WorldScan) that EU air pollution policies will also reduce some GHG emissions, thus the additional cost (carbon price) needed to reduce the remaining GHG emissions could be lower. They assume that 50 percent of the required air pollution emissions reduction will depend on energy efficiency, fuel switching, and structural changes in the economy, and the remaining 50 percent on end-of-pipe controls (with no climate co-benefits). They suggest that because the abatement potential of relatively cheap end-of-pipe options has already been exploited in past decades, further emission reductions through end-of-pipe measures will become more expensive and further reductions of air pollution can be achieved more efficiently through structural changes induced by stricter policies for both air pollution and carbon. Purohit et al. (2010) suggest that an air quality strategy in India using certain climate-friendly measures (such as energy efficiency improvements, fuel substitution, co-generation of heat and power, and integrated coal gasification combined cycle plants) would reduce air pollution at roughly half the cost of an approach that relies on technical end-of-pipe emission-control measures alone. However, Tibrewal and Venkataraman (2020) calculate that, in India, air pollution measures in the domestic sector can have climate-mitigation co-benefits, but reducing air pollution from power plants and transport could have a strong countervailing warming effect by reducing emissions of climate coolants. All three studies found climate co-benefits of air pollution abatement measures by assumption that a country would prioritize climate-friendly measures to achieve air quality goals. As discussed in the next section, such measures are not always the least-cost ways to improve air quality however.

The argument about avoided abatement cost

Climate-led model simulations usually show that climate mitigation reduces the costs of improving air quality. The cost savings come from the observation that climate policies avoid costs of end-of-pipe air pollution measures because they eliminate several sources of combustion of fossil fuels that co-emit air pollutants. Multiple authors argue that a considerable share of investment in climate policies (generally on the order of 20 percent to 50 percent) could be recovered by avoided costs of end-of-pipe air pollution control measures (Brink 2002; McCollum et al. 2013; Rafaj et al. 2013; Rao et al. 2013; Rao et al. 2016; Rao et al. 2017; RIVM et al. 2001; Rogelj, Rao, et al. 2014; Smith 2013; Van Harmelen et al. 2002; Van Vuuren and Bakkes 1999; West et al. 2013). Hamilton et al. (2017) likewise argue that leapfrogging to zero-emission technologies such as renewable energy and some radically energy saving measures has a structural cost advantage compared with fossil fuels with end-of-pipe controls, given that they avoid both pollution mortality and the need for pollution controls.

The argument that climate-mitigation measures reduce the costs of reaching air pollution targets is illustrated in figure 5.5, which was developed in early 2000 for the EU clean air policy by Markus Amann for the International Institute for Applied Systems Analysis (IIASA). It shows that achieving the EU targets of the Kyoto Protocol was expected to reduce the need for air pollution control measures in the European Union by about a quarter (20 billion euros per year), because several fossil fuel combustion sources would have been retired early or not built, so EU member states would not need to put filters on them. The catch is that this argument holds only if a country prioritizes full decarbonization in the first place, is able to pay its price, and is willing to wait longer for air quality benefits delivered by removal of fossil fuel sources rather than early installation of pollution-control technologies on those sources.

FIGURE 5.5

EU air pollution control costs under business-as-usual system versus PRIMES model energy scenario

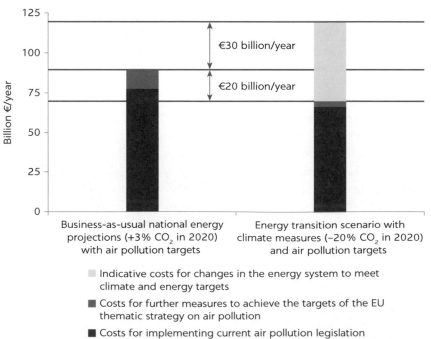

Source: Markus Amann, greenhouse gas–air pollution interactions and synergies (GAINS) model. International Institute for Applied Systems Analysis.
Note: CO_2 = carbon dioxide; EU = European Union.

Climate policies can limit or obviate some local costs of improving air quality but add to the local costs of achieving a global public good. Note that in figure 5.5, the lower cost of air pollution controls for the European Union (€20 billion/year) is more than offset by the added €50 billion/year of the cost of climate mitigation. Consequently, the net increase in the annual combined cost of measures to control air pollution and climate change is about €30 billion/year larger than the costs of air quality measures alone. Advanced economies have a historical obligation confirmed under the Paris Agreement to foot the higher bill of climate mitigation. They also can afford it. However, saving on costs of air pollution by increasing the costs of climate mitigation poses a real-life policy dilemma for the developing countries that struggle with much scarcer resources, lower capabilities, and limited institutional capacity to handle multiple local and global objectives at the same time. If they prioritize costly climate mitigation measures they may have to delay more effective air-pollution control measures with no climate co-benefits. Therefore, models are needed that would provide more relevant insights into the practical sequencing of environmental actions in developing countries.

Toward modeling integrated air pollution and climate change measures

Integrated approaches to study air pollution and climate mitigation incorporate the transmission channels of the impacts of abatement measures that the climate-led models omit, namely the climate penalty of end-of-pipe air pollution control technologies, climate warming by sulfate and nitrate aerosols, and impact of switching from natural gas or fossil fuel-fired electricity to biomass and vice versa. Already in 2010 Van Aardenne et al. noted that although climate change policies do have long-term co-benefits for air quality, by themselves they are not sufficient to solve air pollution problems—additional air pollution policies are also required, especially those that reduce emissions of particulate matter and $PM_{2.5}$ precursors (especially SO_2 and NO_x), which lead to some climate warming. More recent experience and literature reveal that the trade-offs between some air quality management and climate mitigation measures are more common than previously thought. By denying these trade-offs, none of the two environmental challenges can be solved effectively. According to Goldemberg et al. 2018, for example, long-term optimization of costs and benefits of climate mitigation may misguide the prioritization of the air quality improvement measures. Thus, climate policies are not only insufficient to solve air pollution problems but may also be detrimental to air quality through impact channels that climate-led models do not capture. These channels are analyzed in this section and their policy implications are discussed in chapter 6.

The priority abatement measures to improve air quality may be different than those aimed at quick climate mitigation. Back in 2012 Shindell et al. simulated the impact of measures that both mitigate warming and improve air quality, ranking the measures by their climate mitigation potential. They note that if air quality had been a priority in the analysis, the selected measures would be quite different, for example, they would include measures primarily reducing SO_2 emissions, which improve air quality but increase warming. Anenberg et al. (2012) model the potential future global air quality health benefits resulting from implementing 14 specific methane and BC emission-control measures selected for their near-term climate benefits. The results (consistent with Shindell et al. [2012]) suggest that the technological (mostly end-of-pipe) measures for reducing emissions of incomplete combustion, such as BC and its precursors (SO_2 and NO_X), have the highest health benefits, accounting for 72 percent of avoided deaths globally, but their climate benefits are relatively small. The nontechnical measures for BC control that eliminate many polluting activities (sources and fuels) result in much higher climate benefits but result in only 26 percent of avoided deaths from poor air quality. Furthermore, the methane abatement measures, which deliver the largest long-term climate benefits, contribute only about 2 percent of air pollution–related avoided deaths.

Radu et al. (2016) used the IMAGE modeling framework to propose that policies maximizing climate-mitigation effects have the highest impact on reduction of SO_2 and NO_X emissions because of the high overlap of priority sources of emissions of GHGs and air pollutants, whereas their impact on BC and OC emissions is relatively low because the overlap between emission sources is low. The authors found that in most regions, low levels of air pollutant emissions can be achieved by solely implementing stringent air pollution policies, but in Asia and other developing regions, a combination of climate and air pollution policy is needed to bring air pollution levels below those of today. Similar results obtained by Bollen (2015) with the computable general equilibrium model WorldScan. Air pollution policies applied in isolation, let alone isolated climate policies, could not achieve air quality targets. It is worth noting that these results were achieved despite the fact that the impact transmission mechanism that creates co-benefits was still limited to source displacement, as in the climate-driven models.

From now on, we will focus on studies that more realistically analyze complex interactions between climate change mitigation and air pollution abatement measures. The underlying models include the transmission impact channels between air pollution and climate mitigation measures that the climate-led models omit. These include climate penalty of end-of-pipe air pollution control technologies, climate warming by sulfate and nitrate aerosols, and impact of switching from natural gas or fossil fuel–fired electricity to biomass and vice versa. This genre of more integrated modeling approaches indicates that the potential for co-benefits between air pollution and climate-mitigation measures shrinks and for trade-offs increases as the basic, low-cost efficiency improvements have been exploited (Bonilla, Sterner, and Coria 2018; Nam et al. 2014). These studies stress the need for an integrated approach to mitigate carbon emissions and air pollution instead of arguing that air quality will always be a co-benefit of climate mitigation measures.

Ex ante simulations by Portugal-Pereira et al. (2018) for Brazil find that diffusion of low-carbon power generation can have a mixed impact on air pollution. Accelerated phaseout of unabated coal power plants and replacement using plants run on biomass and coal with carbon capture and storage would lead to much lower CO_2 and SO_2 emissions, but emissions of PM and toxic chemicals may rise, indicating the potential for trade-offs. This can happen for two main reasons. First, biomass power plants in the Brazilian context was found to have higher PM emissions than coal per unit of power generated. The introduction of end-of-pipe air pollution control measures on biomass power plants was found to offset some of these trade-offs but comes with additional cost, making coal power plants cost competitive. Second, putting carbon capture and storage facilities on coal power plants would eliminate carbon emissions but would require larger amounts of fuel per kilowatt hour delivered to the grid (the energy penalty discussed earlier in this chapter), which results in higher emissions of particulates and other toxic chemicals when compared with conventional coal technologies without carbon capture and storage.

The recent impact assessment of the EU climate policies (EC 2020), supported by simulations with the Greenhouse Gas–Air Pollution Interactions and Synergies (GAINS) model, illustrates the good practice of jointly addressing air quality and climate change. The baseline scenario for climate policies assumes implementation of aggressive air pollution policies that reduce the combined emissions of SO_2, NO_X, and $PM_{2.5}$ by almost 60 percent by 2030 compared to 2015. Such air pollution abatement requires comprehensive application of end-of-pipe emission controls on most stationary and mobile sources. When additional GHG emissions and carbon pricing constraints are added to these stringent air pollution standards, the model finds another optimal solution with much less fuel use and a larger shift to nonemitting renewable energy sources. This shift would reduce combined emissions of $PM_{2.5}$, NO_X, and SO_2 in the European Union by an additional 3 percent to 10 percent in 2030 relative to the baseline and avoid some costs of end-of-pipe air pollution control. Consequently, the combined application of stringent air pollution controls *and* aggressive decarbonization has more positive impacts on human health than the application of air pollution policies alone. However, the model shows that, in specific locations, air quality may deteriorate (EC 2020).

Vandyck et al. (2018) conducted rare simulations of more-explicit impacts of integrated policy for climate change mitigation and air quality. This study shows that structural changes induced by integrated climate and targeted air quality policy package can improve air quality beyond what can be expected by climate policies or end-of-pipe air pollution abatement technologies alone. However, ramping up the stringency of air quality policies implies massive implementation of end-of-pipe air pollution control technologies that cut emissions of air pollutants that are climate coolants (aerosols) by more than air pollutants that warm the climate (mainly BC), leading to some net warming. (in line with Hienola et al. [2018] and Rogelj, Schaeffer, et al. [2014]). Net warming effect of prioritizing air pollution prevails even before considering the energy, and hence climate penalty of end-of-pipe air pollution controls.

Integrated analyses conducted for countries where residential biomass burning for heating and cooking account for most of the population exposure to $PM_{2.5}$ show another important effect commonly omitted in the climate-led models. The simulations conducted for Serbia's low-carbon development strategy with GEM-E3 model show that PM emissions would initially rise in the low-carbon development scenarios above business as usual as coal-fired district heating and gas used by households for heating gets displaced by biomass, before declining when coal surrenders to natural gas, wind, solar, and hydro power (see figure 5.6). Climate-mitigation measures assumed in the scenario compatible with the EU environmental acquis reduces $PM_{2.5}$ emissions 29 percent below the baseline in 2050. These emissions reductions result from the replacement of fossil fuels across sectors, which reduces CO_2 and associated $PM_{2.5}$ emissions. However, displacement of fossil fuels by biomass burned in small household stoves and boilers reduces GHG emissions but increases emissions of $PM_{2.5}$. Therefore, during the first five years of the implementation of climate policies (between 2020 and 2050), $PM_{2.5}$ emissions increase in the model (figure 5.6). The authors stress that additional air quality management efforts would need to accompany climate policies in the household sector to

FIGURE 5.6

Dynamic impact of Serbia's low-carbon strategy on PM$_{2.5}$ emissions

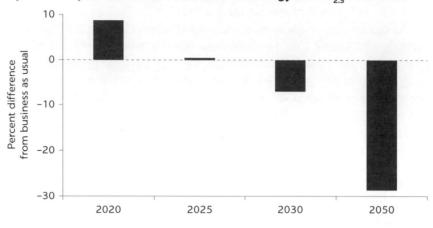

Source: Based on Republic of Serbia (2019).

Note: PM$_{2.5}$ = particulate matter two-and-one-half microns or less in width.

prevent the large-scale rebound of the use of biomass for domestic heating and cooking (Republic of Serbia 2019).

Abatement of short-lived climate pollutants (SLCPs) advances both air quality and climate mitigation but cannot solve either of these problems alone. Already in 2012 Amann identified a program consisting of 16 win-win abatement measures focusing on reducing emissions of SLCPs (see figure 5.7). Amann (2012) is clear that this is not sufficient to improve air quality to heathy standards in most situations. Effective air pollution programs must integrate additional pollution abatement measures that do not have climate co-benefits and reduce conventional air pollutants (especially SO$_2$ and NO$_X$) that cool the climate and slightly increase CO$_2$ emissions (see also Bollen, Guay, et al. 2009; Bollen, van der Zwaan, et al. 2009; Wang et al. 2016). Regarding climate change, a focus on SLCPs, above all methane, could significantly reduce the rate of temperature increase in the next couple of decades and buy more time to implement more aggressive climate policies in the more distant future. However, the SLCP measures alone are far from being able to meet the long-term 2°C goal of the Paris Agreement (let alone the 1.5°C goal) and must be complemented by solutions to contain concentrations of long-lived GHGs, primarily CO$_2$, with much fewer co-benefits for air quality. The IPPC sixth assessment report noted higher climate warming impact of methane than previously assessed and lower of BC (IPPC 2021). Tibrewal and Venkataraman (2020) show that the climate-mitigation co-benefits of air pollution policies in India due to abating SLCP emissions can be more than offset by reducing emissions of climate-cooling agents such as SO$_2$ and NO$_X$. With an understanding of both synergies and trade-offs, policy makers can plan adequate adjustments to long-term climate policies to ensure that air pollution objectives are not compromised.

Few modeling tools can optimize the costs and benefits of comprehensive packages of air pollution and climate-mitigation measures. For example, the GAINS model, developed by IIASA, is a bottom-up technology optimization model with a comprehensive database of about 2,000 technical and nontechnical emission-control measures for 10 air pollutants and 6 GHGs. It uses the marginal abatement potential and cost functions for these measures to find the socially optimum (least cost to society) packages of abatement measures to achieve air quality and climate objectives, capturing many real-life

FIGURE 5.7

Selected air quality measures with co-benefits to climate change identified by the International Institute for Applied Systems Analysis

Methane measures	Technical black carbon measures	Nontechnical measures
• Recovery of coal mine gas • Reduction of gas venting associated with production of crude oil and natural gas • Reduction of gas leakages from pipelines and distribution nets • Waste recycling • Wastewater treatment • Farm-scale anaerobic digestion • Aeration of rice paddies	• Modern coke ovens • Modern brick kilns • Diesel particle filters • Briquettes instead of coal for heating • Improved biomass cookstoves • Pellet stoves and boilers (in industrial countries)	• Ban of high-emitting vehicles • Ban of open burning of agricultural waste • Elimination of biomass cook stoves

Source: Amann 2012.

synergies and trade-offs between different abatement measures that other models leave out. The recent application of the GAINS model in Kazakhstan illustrated that while several low-cost air pollution abatement measures demonstrate potential for also reducing GHG emissions, some measures such as improving environmental performance of industrial processes and installing end-of-the pipe SO_2 and NO_x control technologies on coal power plants would reverse a portion of mitigation gains of win-win measures (figure 5.8). The climate co-benefits of cost-effective air quality improvement measures are more common in Kazakhstan, where the majority of population exposure to $PM_{2.5}$ comes from combustion of coal in inefficient stoves and boilers in residential buildings during winter heating season and massive coal combustion in power, industrial, and heating plants located close to city centers. Kazakhstan households use very little biomass unlike other emerging economies in Central and Eastern Europe (including Serbia shown in figure 5.6) and many developing countries in warmer climate where biomass is a dominant household fuel for cooking and water heating.

Each dot in figure 5.8 moving from right to left on both schedules represents a modeled measure to improve air quality (orange curve) and its respective impact on GHG emissions (red curve). The down-sloping red curve means an air pollution measure also reduces GHG emissions (synergy), while the upward-sloping red curve implies a trade-off. The figure shows that reducing mean national exposure from 28 µg/m³ to 18.5 µg/m³ can have a number of climate co-benefits. The measures showing the largest potential for cost-effective reduction of both mean $PM_{2.5}$ population exposure and GHG emissions are (1) replacing individual coal stoves and boilers with connections to improved district heating and conversion to natural gas or liquefied petroleum gas (LPG), briquettes, or heat pumps; (2) improving building energy efficiency; and (3) improving waste management. The main trade-off is associated with installation of emission control equipment at power plants (highlighted in blue). This would bring population exposure closer to 18 µg/m³ of $PM_{2.5}$. The switch from biomass to gas/LPG was not included among available abatement measures because the use of biomass for residential heating and cooking is minimal in Kazakhstan's cities.

FIGURE 5.8

Marginal cost curve for reducing population exposure to PM₂.₅ in Kazakhstan in 2030 and the impacts of the air pollution abatement measures on GHG emissions

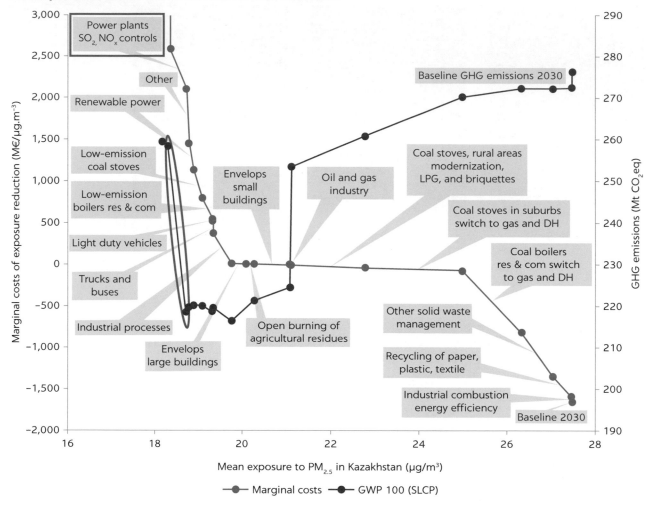

Source: Zlatev et al. 2021.

Notes: GHG emissions are expressed as Mt of CO_2 equivalent using the Global Warming Potential over 100 years (GWP100) metric that accounts for short-lived climate pollutants (SLCPs) impacts over a 100-year time horizon. The orange curve represents a marginal air pollution abatement cost curve for Kazakhstan with a potential to reduce mean exposure to PM₂.₅ on the horizontal axis and a cost of reducing exposure by one unit on the vertical left axis. The red schedule represents the impact of air pollution technical abatement measures on national GHG emissions (right) axis). The curves should be read from right to left. The first dot to the right on the orange curve represents projected baseline PM₂.₅ mean exposure in Kazakhstan in 2030, whereas the first dot to the right on the red curve represents projected baseline GHG emissions in Kazakhstan in 2030. GHG = greenhouse gas; LPG = liquefied petroleum gas; NOₓ = nitrogen oxide; PM₂.₅ = particulate matter two-and-one-half microns or less in width; SO₂ = sulfur dioxide; µg/m³ = micrograms per cubic meter.

Energy efficiency can make cost-effective strides toward better air quality, but it is not a low-hanging fruit as most models (including GAINS) assume. Models often underestimate the costs and overestimate the sustainable emissions-reduction potential of energy efficiency. They calculate the costs and benefits from the perspective of a social planner rather than from the perspective of real economic agents. In reality energy efficiency projects require that parts of the buildings on production units be taken out of operation for a period; change in working practices is disruptive; new habits must be learned; senior managers have many other things to worry about; incentives are split between the owners and users of buildings and equipment, and so on. These transaction costs are not visible in most traditional models. Neither are the specific risks, cost of capital, and cost of acquiring information that economic actors face. Social planners are oblivious to specific risks and are very patient. The models representing social planners' perspective optimize for the society in a very long term, using low social discount rates. On the benefit side of the cost-benefit equation social planners also miss a lot of crucial variables, such as which air pollution sources are first decommissioned due to lower energy demand. The analyses of energy efficiency potential also rarely consider the rebound effect—after implementation of efficiency measures that make a unit of energy service cheaper (for example get more light for unit of electricity used), users often begin to increase luminosity, building temperature, and kilometers driven. (Aydin, Kok, and Brounen 2017; Azevedo 2014; Sorrell, Gatersleben, and Druckman 2020). The rebound effect offset between 25 percent and more than 100 percent of the technical energy saving potential of projects, and is stronger when energy-efficiency projects are subsidized but energy prices remain low and even subsidized. The first model that aims at representing the investment and behavioral perspective of firms and households operating in specific markets in Bulgaria and Croatia has been presented in Vasquez-Suarez, Liu, and Peszko 2018), and later extended to support the World Bank energy efficiency policy dialogue in Morocco. Putting the correct policy incentives in place can greatly enhance the probability that low-cost improvements will be adopted and rebound effect prevented. The largest and most permanent improvements in energy efficiency have been achieved in the former communist countries as part of deep, market-based structural transformation of wasteful centrally planned industry and infrastructure. But in present-day China, or Central and Eastern Europe, the cheapest energy-saving opportunities have been exhausted and expected market rates of return are much higher. Therefore, the marginal costs of energy efficiency come closer to the costs of other abatement measures and the marginal benefits of energy efficiency are shrinking.

Limiting air quality measures to those with climate co-benefits may increase the overall costs of air pollution control, as demonstrated by Cameron et al. (2016) in their analysis of the proliferation of modern clean cookstoves in South Asia. Their stringent climate-mitigation scenario increases clean fuel (LPG and LNG) costs for households by 38 percent in 2030 relative to the baseline, leading to 21 percent more South Asians using traditional polluting stoves. To achieve the goal of increasing universal access to clean cooking nonetheless, government subsidies would need to increase by 44 percent because clean cooking alternatives with climate co-benefits (renewable electricity, improved biomass, biofuels) are more expensive than LPG or LNG stoves. The authors argue that these additional costs of ensuring climate co-benefits of clean cooking can be offset by additional external concessional financing. However, these financial transfers would need to materialize. As mentioned above, climate finance is currently rarely available for conversion of biomass stoves to clean cooking using LNG or LPG as a switch from renewable energy to fossil fuels.

The traditional models discussed in this section suggest that a world without fossil fuels will be a world with cleaner air and healthier people. This section discussed three major omissions in the typical modeling framework that limits applicability of this hypothesis. Furthermore, such a world is decades and billions of dollars away. In the meantime, every year more than 7 million people die from air pollution globally. Solutions to air pollution crises must be local and implemented in incremental steps. High-income countries can afford to leap-frog to a decarbonized economy relatively quickly and obtain air quality as a co-benefit. In lower-income countries, however, limited resources and weak institutional capacity force local policy makers, especially in developing countries, to prioritize measures in the short term to prevent as many premature deaths as quickly as possible at reasonable cost. At the same time, however, decision-makers should minimize large irreversible investments in carbon-intensive technologies and infrastructure that can create costly future liability. These are the daily policy dilemmas in polluted cities discussed in chapter 6. For example, a study on air pollution in Asia and the Pacific (CCAC and UNEP 2019) identifies 25 clean measures with a potential to reduce the number of people exposed to pollution by 80 percent, even though some individual measures would warm the climate. Therefore, the sequencing of these measures under hard budget constraints would be different depending on whether air quality or climate was the priority.

Despite the synergies of implementing all measures in the long run, the short-term trade-offs are part of life. Unlike climate change, which is caused by a *stock* of GHGs slowly accumulating in the atmosphere, air pollution is a *flow* phenomenon—the benefits of avoided deaths and diseases are realized almost immediately after emissions have been reduced. Therefore, the priority in air quality management is to reduce population exposure in the most polluted areas as quickly as possible. Switching to cleaner fossil fuels for household cooking and heating, moving traffic congestion out of the city center, and relocating the worst sources of pollution and retrofitting them with pollution-control technologies are among the measures that will rapidly improve local pollution but may not make much difference to climate change. Deep decarbonization of energy, transport, and industrial systems takes decades to achieve and cannot be controlled by local policy makers alone. Optimizing the cost of climate mitigation often implies that some air quality improvements would not be implemented at all or would be applied later and at smaller scale. The typical measures that are limited by the climate-led agenda is switching from biomass to natural gas in household heating and cooking or retrofitting combustion plants with filters and scrubbers. These measures can quickly save most lives from air pollution but require additional climate action elsewhere to compensate for their small climate penalty.

This chapter shows that even though air and climate pollutants are often co-emitted from the same sources, the same abatement measures may either deliver a double environmental dividend for air pollution and for climate or lead to trade-offs between these two environmental problems. Awareness of these complexities and capturing them in technical and economic models helps prepare integrated policy programs to mitigate air pollution and climate change and thereby harness synergies, but also manage inevitable short-term and medium-term trade-offs. Such integrated policies prioritize air pollution measures where appropriate, while remaining flexible and paving the way for long-term decarbonization of economic activities. This is discussed in chapter 6.

NOTES

1. The best-known marginal abatement cost model for calculating cost-effectiveness of air pollution programs is GAINS by the International Institute for Applied Systems Analysis.
2. "Diesel Engine: How Does the DPF Effects on Performance of Engine?" (https://www.researchgate.net/post/Diesel_Engine_How_does_the_DPF_effects_on_performance_of_engine).
3. "Statement of the Manufacturers of Emission Controls Association on the U.S. Environmental Protection Agency's Proposed Rulemaking on Greenhouse Gas Emissions Standards and Fuel Efficiency Standards for Medium- and Heavyduty Engines and Vehicles – Phase 2" (http://www.meca.org/attachments/2674/MECA_comments_on_EPA_Phase_2_HD_GHG_092515_final.pdf).
4. Input provided by Gordon Hughes, Professorial Fellow, School of Economics, University of Edinburgh.

REFERENCES

Agee, M. D., S. E. Atkinson, T. D. Crocker, and J. W. Williams. 2014. "Non-Separable Pollution Control: Implications for a CO_2 Emissions Cap and Trade System." *Resource and Energy Economics* 36 (1): 64–82. https://doi.org/10.1016/j.reseneeco.2013.11.002.

Amann, Markus. 2012. "Modelling Co-Benefits with IIASA's GAINS Model: Key Findings." Paper presented at the International Workshop on a Co-benefit Approach, Institute for Global Environmental Strategies, February 13–14, 2012. https://archive.iges.or.jp/en/archive/cp/pdf/activity20120213/s3-2_amann.pdf.

Andaloussi, Mehdi Benatiya. 2018. "Clearing the Air: The Role of Technology Adoption in the Electricity Generation Sector." Working Paper, Toulouse School of Economics.

Anenberg, Susan C., Joel Schwartz, Drew Shindell, Markus Amann, Greg Faluvegi, Zbigniew Klimont, Greet Janssens-Maenhout, et al. 2012. "Global Air Quality and Health Co-Benefits of Mitigating Near-Term Climate Change through Methane and Black Carbon Emission Controls." *Environmental Health Perspectives* 120 (6): 831–39. doi:10.1289/ehp.1104301.

ARB (Air Resources Board). 2014. California Air Resources Board's Clearinghouse of Non-CO_2 Greenhouse Gas Emission Control Technologies. https://ww3.arb.ca.gov/cc/non-co2-clearinghouse/technology/b-2-1.pdf.

Aydin, Erdal, Nils Kok, and Dirk Brounen. 2017. "Energy Efficiency and Household Behavior: The Rebound Effect in the Residential Sector." *RAND Journal of Economics* 48 (3). https://doi.org/10.1111/1756-2171.12190.

Azevedo, Inês M. L. 2014. "Consumer End-Use Energy Efficiency and Rebound Effects." *Annual Review of Environment and Resources* 39 (1): 393–418. doi:10.1146/annurev-environ-021913-153558.

Bollen, J. C. 2015. "The Value of Air Pollution Co-Benefits of Climate Policies: Analysis with a Global Sector-Trade CGE Model Called WorldScan." *Technological Forecasting and Social Change* 90: 178–91.

Bollen, J., and C. Brink. 2012. "Air Pollution Policy in Europe: Quantifying the Interaction with Greenhouse Gases and Climate Change Policies." CPB Discussion Paper, CPB Netherlands Bureau for Economic Policy Analysis, The Hague.

Bollen, J. C., Bruno Guay, Stéphanie Jamet, and Jan Corfee-Morlot. 2009. "Co-Benefits of Climate Change Mitigation Policies: Literature Review and New Results." Economics Department Working Paper 693, OECD, Paris. https://www.oecd-ilibrary.org/economics/co-benefits-of-climate-change-mitigation-policies_224388684356.

Bollen, J. C., B. van der Zwaan, C. Brink, and H. Eerens. 2009. "Local Air Pollution and Global Climate Change: A Combined Cost-Benefit Analysis." *Resource and Energy Economics* 31 (3): 161–81. https://econpapers.repec.org/article/eeeresene/v_3a31_3ay_3a2009_3ai_3a3_3ap_3a161-181.htm.

Bonilla, Jorge Londoño, Jessica Coria, and Thomas Sterner. 2018. "Technical Synergies and Trade-Offs between Abatement of Global and Local Air Pollution." *Environmental & Resource Economics* 70 (1): 191–221.

Brännlund, R., and B. Kriström. 2001. "Too Hot to Handle? Benefits and Costs of Stimulating the Use of Biofuels in the Swedish Heating Sector." *Resource Energy Economics* 23: 343–58.

Brink, C. 2002. *Modelling Cost-Effectiveness of Interrelated Emission Reduction Strategies: The Case of Agriculture in Europe.* PhD Thesis, Wageningen University.

Bruce, Nigel, Kristin Aunan, and Eva A. Rehfuess. 2017. "Liquefied Petroleum Gas as a Clean Cooking Fuel for Developing Countries: Implications for Climate, Forests, and Affordability." KfW Development Bank, Frankfurt, Germany. https://www.kfw-entwicklungsbank.de /PDF/Download-Center/Materialien/2017_Nr.7_CleanCooking_Lang.pdf.

Burnard, Keith, Julie Jiang, Bo Li, Gilbert Brunet, and Franz Bauer. 2014. "Emissions Reduction through Upgrade of Coal-Fired Power Plants: Learning from Chinese Experience." International Energy Agency, Paris.

Cameron, Colin, Shonali Pachauri, Narasimha D. Rao, David McCollum, Joeri Rogelj, and Keywan Riahi. 2016. "Policy Trade-Offs between Climate Mitigation and Clean Cook-Stove Access in South Asia." *Nature Energy* 1: 15010. https://doi.org/10.1038/nenergy.2015.10.

Carrington, Damian. 2016. "More than 1,000 Diesel Cars Caught without Pollution Filter, Figures Show." *The Guardian*, April 17, 2016. https://www.theguardian.com/environment /2016/apr/17/diesel-particulate-filter-removal-air-pollution-department-for-transport.

CCAC (Climate and Clean Air Coalition) and UNEP (United Nations Environment Programme). 2019. *Air Pollution in Asia and the Pacific: Science-Based Solutions.* Asia Pacific Clean Air Partnership. Nairobi: UNEP.

Coady, David, Ian W. H. Parry, Nghia-Piotr Le, and Baoping Shang. 2019. "Global Fossil Fuel Subsidies Remain Large: An Update Based on Country-Level Estimates." Working Paper 19/89, International Monetary Fund, Washington, DC.

Coady, David, Ian Parry, Louis Sears, and Baoping Shang. 2017. "How Large Are Global Fossil Fuel Subsidies?" *World Development* 91 (March): 11–27. https://doi.org/10.1016/j .worlddev.2016.10.004.

Cropper, Maureen L., Sarath Guttikunda, Puja Jawahar, Kabir Malik, and Ian Partridge. 2017. "Costs and Benefits of Installing Flue-Gas Desulfurization Units at Coal-Fired Power Plants in India." In *Disease Control Priorities*, 3rd edition, Volume 7 *Injury Prevention and Environmental Health*, chapter 13. Washington, DC: International Bank for Reconstruction and Development / World Bank. https://www.ncbi.nlm.nih.gov/books/NBK525204 /doi:10.1596/978-1-4648-0522-6/ch13.

Dulal, H. B., and S. Akbar. 2013. "Greenhouse Gas Emission Reduction Options for Cities: Finding the 'Coincidence of Agendas' between Local Priorities and Climate Change Mitigation Objectives." *Habitat International* 38: 100–5. doi:10.1016/j.habitatint.2012.05.001.

EC (European Commission). 2020. "Stepping Up Europe's 2030 Climate Ambition: Investing in a Climate-Neutral Future for the Benefit of Our People." Commission Staff Working Document Impact Assessment Accompanying the document Communication from the Commission to the European Parliament, the Council, The European Economic and Social Committee and the Committee of the Regions, Brussels, 17.9.2020, SWD(2020) 176 final. https://eur-lex.europa.eu/resource.html?uri=cellar:749e04bb-f8c5-11ea-991b-01aa 75ed71a1.0001.02/DOC_1&format=PDF.

Edwards, Robert, Heinz Hass, Jean-François Larivé, Laura Lonza, Heiko Maas, and David Rickeard. 2014. "Well-to-Wheels Analysis of Future Automotive Fuels and Powertrains in the European Context." Joint Research Center Technical Report, European Commission, Brussels.

EIB (European Investment Bank). 2020. "Project Carbon Footprint Methodologies. Methodologies for the Assessment of Project GHG Emissions and Emission Variations." EIB, Luxembourg. https://www.eib.org/attachments/strategies/eib_project_carbon _footprint_methodologies_en.pdf.

EPRI (Electric Power Research Institute). 2011. *Electricity Use in the Electric Sector.* Palo Alto, CA: EPRI. https://www.epri.com/research/products/000000000001024651.

ESMAP (Energy Sector Management Assistance Program). 2020. "Quantifying and Measuring Climate, Health, and Gender Co-Benefits from Clean Cooking Interventions: Methodologies Review." World Bank, Washington, DC.

European Environmental Agency (EEA). 2019. "Assessing the Effectiveness of EU Policy on Large Combustion Plants in Reducing Air Pollutant Emissions." https://www.eea.europa.eu/publications/effectiveness-of-eu-policy-on.

Forbes, Alex. 2018. "Cleaning up with FGD Technology: Making Profitable Trade-Offs in Desulfurizing Power." https://www.ge.com/power/transform/article.transform.articles.2018.jan.cleaning-up-with-fgd-technolog.

Fujii, Hidemichi, Shunsuke Managi, and Shinji Kaneko. 2013. "Decomposition Analysis of Air Pollution Abatement in China: Empirical Study for Ten Industrial Sectors from 1998 to 2009." *Journal of Cleaner Production* 59: 22–31. http://dx.doi.org/10.1016/j.jclepro.2013.06.059.

GACC (Global Alliance for Clean Cooking). 2014. *Clean Cookstoves and Fuels: A Catalog of Carbon Offset Projects and Advisory Service Providers,* 2nd edition. Washington, DC: GACC. https://www.cleancookingalliance.org/binary-data/RESOURCE/file/000/000/381-1.pdf.

GAHP (Global Alliance on Health and Pollution), AirQualityAsia, and the Schiller Institute for Integrated Science and Society at Boston College. 2020. "Air Pollution Interventions: Seeking the Intersection between Climate and Health." GAHP, New York. https://gahp.net/report-air-pollution-interventions-seeking-the-intersection-between-climate-health/.

Goldemberg, Jose, Javier Martinez-Gomez, Ambuj Sagar, and Kirk R. Smith. 2018. "Household Air Pollution, Health, and Climate Change: Cleaning the Air." *Environmental Research Letters* 13 (030201). https://doi.org/10.1088/1748-9326/aaa49d.

Government of India. 2010. *Report of the Expert Group on a Viable and Sustainable System of Pricing of Petroleum Products.* New Delhi. February 2, 2010. http://www.indiaenvironmentportal.org.in/files/reportprice.pdf.

Hamilton, Ian, Harry Kennard, Alice McGushin, Lena Höglund-Isaksson, Gregor Kiesewetter, Melissa Lott, James Milner, et al. 2021. "The Public Health Implications of the Paris Agreement: A Modelling Study." *Lancet Planet Health* 5: e74–83.

Hamilton, Kirk, Milan Brahmbhatt, Nicholas Bianco, and Jiemei Liu. 2017. "Multiple Benefits from Climate Mitigation: Assessing the Evidence." Grantham Research Institute on Climate Change and the Environment, London School of Economics, London. https://www.researchgate.net/publication/320867824_Multiple_Benefits_from_Climate_Change_Mitigation_Assessing_the_Evidence.

Heger, Martin, and Maria Sarraf. 2018. "Air Pollution in Tehran: Health Costs, Sources, and Policies." Environment and Natural Resources Global Practice, Discussion Paper 6, World Bank, Washington, DC. https://openknowledge.worldbank.org/handle/10986/29909.

Hienola, Anca, Antti-Ilari Partanen, Joni-Pekka Pietikäinen, Declan O'Donnell, Hannele Korhonen, Damon Matthews, and Ari Laaksonen. 2018. "The Impact of Aerosol Emissions on the 1.5°C Pathways." *Environmental Research Letters* 13 (4) 044011. https://iopscience.iop.org/article/10.1088/1748-9326/aab1b2/meta.

Huppmann, D., E. Kriegler, V. Krey, K. Riahi, J. Rogelj, S. K. Rose, J. Weyant, et al. 2019. "IAMC 1.5°C Scenario Explorer and Data." Paper presented at meeting of the International Institute for Applied Systems Analysis. Integrated Assessment Modeling Consortium and International Institute for Applied Systems Analysis. data.ene.iiasa.ac.at/iamc-1.5c-explorer.

ICCT (International Council on Clean Transportation). 2016. "A Technical Summary of Euro 6/VI Vehicle Emission Standards." ICCT Briefing. https://theicct.org/sites/default/files/publications/ICCT_Euro6-VI_briefing_jun2016.pdf.

IEA (International Energy Agency). 2017. *Energy Access Outlook 2017: From Poverty to Prosperity.* Paris: IEA. https://webstore.iea.org/weo-2017-special-report-energy-access-outlook.

IER (Institute for Energy Research). 2017. "Cleaned-Up Coal and Clean Air: Facts about Air Quality and Coal-Fired Power Plants." IER, Washington, DC. https://www.instituteforenergyresearch.org/uncategorized/cleaned-coal-clean-air-facts-air-quality-coal-fired-power-plants/.

IMF (International Monetary Fund). 2019. *Fiscal Monitor, October 2019: How to Mitigate Climate Change.* Washington, DC: IMF.

IPCC. 2021. "Summary for Policymakers". In: *Climate Change 2021: The Physical Science Basis. Contribution of Working Group I to the Sixth Assessment Report of the Intergovernmental Panel on Climate Change* [Masson-Delmotte, V., P. Zhai, A. Pirani, S.L. Connors, C. Péan, S. Berger, N. Caud, Y. Chen, L. Goldfarb, M.I. Gomis, M. Huang, K. Leitzell, E. Lonnoy, J.B.R. Matthews, T.K. Maycock, T. Waterfield, O. Yelekçi, R. Yu, and B. Zhou (eds.)]. In Press.

Johnson, M., R. Edwards, and O. Masera. 2010. "Improved Stove Programs Need Robust Methods to Estimate Carbon Offsets." *Climatic Change* 102 (3–4): 641–49. doi:10.1007/s10584-010-9802-0.

Khreis, H., A. D. May, and M. J. Nieuwenhuijsen. 2017. "Health Impacts of Urban Transport Policy Measures: A Guidance Note for Practice." *Journal of Transport and Health* 6: 209–27. https://doi.org/10.1016/j.jth.2017.06.003.

Kypridemos, Chris, Elisa Puzzolo, Borgar Aamaas, Lirije Hyseni, Matthew Shupler, Kristin Aunan, and Daniel Pope. 2020. "Health and Climate Impacts of Scaling Adoption of Liquefied Petroleum Gas (LPG) for Clean Household Cooking in Cameroon: A Modeling Study." *Environmental Health Perspectives* 128 (4): 47001. doi:10.1289/EHP4899.

Lee, Carrie M., Chelsea Chandler, Michael Lazarus, and Francis X. Johnson. 2013. "Assessing the Climate Impacts of Cookstove Projects: Issues in Emissions Accounting." Working Paper 2013-01, Stockholm Environment Institute.

Letnik, T., M. Marksel, G. Luppino, A. Bardi, and S. Božičnik. 2018. "Review of Policies and Measures for Sustainable and Energy Efficient Urban Transport." *Energy* 163: 245–57. https://doi.org/10.1016/j.energy.2018.08.096.

Li, M., D. Zhang, C. T. T. Li, K. M. Mulvaney, N. E. Selin, and V. J. Karplus. 2018. "Air Quality Co-Benefits of Carbon Pricing in China." *Nature Climate Change* 8: 398. https://doi.org/10.1038/s41558-018-0139-4.

Lyu, Wanning, Yuan Li, Dabo Guan, Hongyan Zhao, Qiang Zhang, and Zhu Liu. 2016. "Driving Forces of Chinese Primary Air Pollution Emissions: An Index Decomposition Analysis." *Journal of Cleaner Production* 133. 10.1016/j.jclepro.2016.04.093.

Markandya, Anil, Ben G. Armstrong, Simon Hales, Aline Chiabai, Patrick Criqui, Silvana Mima, Cathryn Tonne, and Paul Wilkinson. 2009. "Public Health Benefits of Strategies to Reduce Greenhouse-Gas Emissions: Low-Carbon Electricity Generation." *Lancet* 374 (9706): 2006–15.

Markandya, Anil, Jon Sampedro, Steven J. Smith, Rita Van Dingenen, Cristina Pizarro-Irizar, Iñaki Arto, and Mikel González-Eguino. 2018. "Health Co-Benefits from Air Pollution and Mitigation Costs of the Paris Agreement: A Modelling Study." *Lancet Planet Health* 2: e126–33.

Massetti, Emanuele, Marilyn A. Brown, Melissa Lapsa, Isha Sharma, James Bradbury, Colin Cunliff, and Yufei Li. 2017. *Environmental Quality and the U.S. Power Sector: Air Quality, Water Quality, Land Use and Environmental Justice*. Oak Ridge, TN: Oak Ridge National Laboratory. https://www.energy.gov/sites/prod/files/2017/01/f34/Environment%20Baseline%20Vol.%202--Environmental%20Quality%20and%20the%20U.S.%20Power%20Sector--Air%20Quality%2C%20Water%20Quality%2C%20Land%20Use%2C%20and%20Environmental%20Justice.pdf.

Masters, Gilbert M. 2004. *Renewable and Efficient Electric Power Systems*. Hoboken, NJ: Wiley-Interscience. https://onlinelibrary.wiley.com/doi/book/10.1002/0471668826.

McCollum, D. L., V. Krey, P. Riahi, P. Kolp, A. Grubler, M. Makowski, and N. Nakicenovic. 2013. "Climate Policies Can Help Resolve Energy Security and Air Pollution Challenges." *Climatic Change* 119: 479–94.

Nam, K. M., C. J. Waugh, S. Paltsev, J. Reilly, and V. Karplus. 2014. "Synergy between Pollution and Carbon Emissions Control: Comparing China and the United States." *Energy Economics* 46: 186–201.

Nemet, G. F., T. Holloway, and P. Meier. 2010. "Implications of Incorporating Air-Quality Co-Benefits into Climate Change Policymaking." *Environmental Research Letters* 5: 014007.

OECD (Organisation for Economic Co-operation and Development). 2019. *Taxing Energy Use 2019: Using Taxes for Climate Action.* Paris: OECD Publishing. https://doi .org/10.1787/058ca239-en.

OECD (Organisation for Economic Co-operation and Development). 2020. Best Available Techniques (BAT) for Preventing and Controlling Industrial Pollution, Activity 4: Guidance Document on Determining BAT, BAT Associated Environmental Performance Levels and BAT-Based Permit Conditions, Environment, Health and Safety, Environment Directorate, OECD

Pachauri, Shonali, Narasimha D. Rao, and Colin Cameron. 2018. "Outlook for Modern Cooking Energy Access in Central America." *PLoS One* 13 (6): e0197974. doi:10.1371/journal .pone.0197974.

Parry, Ian, Dirk Heine, Eliza Lis, and Shanjun Li. 2014. *Getting Energy Prices Right: From Principles to Practice.* Washington, DC: International Monetary Fund.

Peszko, Grzegorz, Dominique van der Mensbrugghe, and Alexander Golub. 2020. "Diversification and Cooperation Strategies in a Decarbonizing World." Policy Research Working Paper 9315, World Bank, Washington, DC. https://openknowledge.worldbank.org/handle /10986/34056.

Pittel, K., and D. Rübbelke. 2008. "Climate Policy and Ancillary Benefits: A Survey and Integration into the Modelling of International Negotiations on Climate Change." *Ecological Economics* 68 (1–2): 210–20.

Portugal-Pereira, Alexandre Koberle, André F. P. Lucena, Pedro R. R. Rochedo, Mariana Império, Ana Monteiro Carsalade, Roberto Schaeffer, and Peter Rafaj. 2018. "Interactions between Global Climate Change Strategies and Local Air Pollution: Lessons Learnt from the Expansion of the Power Sector in Brazil." *Climatic Change* 148: 293–309. https://doi .org/10.1007/s10584-018-2193-3.

Poullikkas, Andreas. 2015. "Review of Design, Operating, and Financial Considerations in Flue Gas Desulfurization Systems." *Energy Technology & Policy* 2 (1): 92–103. doi:10.1080/233170 00.2015.1064794.

Purohit, Pallav, Markus Amann, Ritu Mathur, Ila Gupta, Sakshi Marwah, Vishal Verma, Imrich Bertok, et al. 2010. *GAINS Asia: Scenarios for Cost-Effective Control of Air Pollution and Greenhouse Gases in India.* Laxenburg, Austria: GAINS International Institute for Applied Systems Analysis (IIASA).

Radu, O. B., M. van den Berg, Z. Klimont, S. Deetman, G. Janssens-Maenhout, M. Muntean, C. Heyes, F. Dentener, and D. P. van Vuuren. 2016. "Exploring Synergies between Climate and Air Quality Policies Using Long-Term Global and Regional Emission Scenarios." *Atmospheric Environment.* 140 (C): 577–91. doi:10.1016/j.atmosenv .2016.05.021.

Rafaj, Peter, and Markus Amann. 2018. "Decomposing Air Pollutant Emissions in Asia: Determinants and Projections." *Energies* 11: 1299. doi:10.3390/en11051299.

Rafaj, Peter, Markus Amann, and José G. Siri. 2014. "Factorization of Air Pollutant Emissions: Projections versus Observed Trends in Europe." *Science of the Total Environment* 494–495: 272–82. doi:10.1016/j.scitotenv.2014.07.013.

Rafaj, Peter, Markus Amann, José G. Siri, and Henning Wuester. 2014. "Changes in European Greenhouse Gas and Air Pollutant Emissions 1960–2010: Decomposition of Determining Factors." *Climatic Change* 124: 477–504.

Rafaj, P., W. Schöpp, P. Russ, C. Heyes, and M. Amann. 2013. "Co-Benefits of Post-2012 Global Climate Mitigation Policies." *Mitigation and Adaptation Strategies for Global Change* 18: 801–24. https://doi.org/10.1007/s11027-012-9390-6.

Rao, Shilpa, Zbigniew Klimont, Joana Leitao, Keywan Riahi, Rita van Dingenen, Lara Aleluia Reis, Katherine Calvin, et al. 2016. "A Multi-Model Assessment of the Co-Benefits of Climate Mitigation for Global Air Quality." *Environmental Research Letters* 11 (12).

Rao, S., Z. Klimont, S. J. Smith, R. Van Dingenen, F. Dentener, L. Bouwman, K. Riahi, et al. 2017. "Future Air Pollution in the Shared Socio-Economic Pathways." *Global Environmental Change* 42: 346–58.

Rao, S., S. Pachauri, F. Dentener, P. Kinney, Z. Klimont, K. Riahi, and W. Schoepp. 2013. "Better Air for Better Health: Forging Synergies in Policies for Energy Access, Climate Change and Air Pollution." *Global Environmental Change* 23: 1122–30.

Rauner, S., J. Hilaire, D. Klein, J. Strefler, and G. Luderer. 2020. "Air Quality Co-Benefits of Ratcheting up the NDCs." *Climatic Change* 163: 1481–500. https://doi.org/10.1007/s10584 -020-02699-1.

Republic of Serbia. 2019. "Draft Low Carbon Development Strategy with Action Plan." GFA Consulting Group GmbH, Hamburg. http://www.serbiaclimatestrategy.eu/.

Richards, John R. 2000. *Control of Nitrogen Oxides Emissions. Student Manual.* APTI Course 418. United States Environmental Protection Agency. http://www.4cleanair.org /APTI/418Combinedchapters.pdf.

RIVM (Rijksinstituut voor Volksgezondheid en Mileu [Dutch National Institute for Health Environment]), EFTEC (Economics for the Environment Consultancy), NTUA (National Technical University of Athens), and IIASA (International Institute for Applied Systems Analysis). 2001. *European Environmental Priorities: An Integrated Economic and Environmental Assessment.* Bilthoven, Netherlands: National Institute of Public Health and the Environment. https://www.rivm.nl/bibliotheek/rapporten/481505010.pdf.

Rogelj, J., D. Huppmann, V. Krey, K. Riahi, L. Clarke, M. Gidden, Z. Nicholls, and M. Meinshausen. 2019. "A New Scenario Logic for the Paris Agreement Long-Term Temperature Goal." *Nature* 573: 357–63. https://doi.org/10.1038/s41586-019-1541-4.

Rogelj, J., A. Pop., K. V. Calvin, G. Luderer, J. Emmerling, D. Gernaat, S. Fujimori, et al. 2018. "Scenarios towards Limiting Global Mean Temperature Increase below 1.5 °C." *Nature Climate Change* 8: 325–32. https://doi.org/10.1038/s41558-018-0091-3, with supplementary materials. https://static-content.springer.com/esm/art%3A10.1038%2Fs41558-018-0091-3 /MediaObjects/41558_2018_91_MOESM1_ESM.pdf.

Rogelj, J., S. Rao, D. L. McCollum, S. Pachauri, Z. Klimont, V. Krey, and K. Riahi. 2014. "Air-Pollution Emission Ranges Consistent with the Representative Concentration Pathways." *Nature Climate Change* 4: 446–50. doi:10.1038/nclimate2178.

Rogelj, J., M. Schaeffer, M. Meinshausen, D. T. Shindell, W. Hare, Z. Klimont, G. J. M. Velders, M. Amann, and H. J. Schellnhuber. 2014. "Disentangling the Effects of CO_2 and Short-Lived Climate Forcer Mitigation." *Proceedings of the National Academy of Sciences of the United States of America* 111 (46): 16325–30. doi:10.1073/pnas.1415631111.

Rogelj, J., D. Shindell, K. Jiang, S. Fifita, P. Forster, V. Ginzburg, C. Handa, et al. 2018. "Mitigation Pathways Compatible with 1.5°C in the Context of Sustainable Development." In *Global Warming of 1.5°C: An IPCC Special Report on the Impacts of Global Warming of 1.5°C above Pre-Industrial Levels and Related Global Greenhouse Gas Emission Pathways, in the Context of Strengthening the Global Response to the Threat of Climate Change, Sustainable Development, and Efforts to Eradicate Poverty*, edited by V. Masson-Delmotte, P. Zhai, H.-O. Pörtner, D. Roberts, J. Skea, P.R. Shukla, A. Pirani, et al. Geneva: Intergovernmental Panel on Climate Change.

Rosenthal, J., A. Quinn, A. P. Grieshop, A. Pillarisetti, and R. I. Glass. 2018. "Clean Cooking and the SDGs: Integrated Analytical Approaches to Guide Energy Interventions for Health and Environment Goals." *Energy for Sustainable Development* 42: 152–59.

Rubin, Edward S., and Due G. Nguyen. 1978. "Energy Requirements of a Limestone FGD System." *Journal of the Air Pollution Control Association* 28 (12): 1207–12. doi:10.1080/00022 470.1978.10470728.

Ryani, Lisa, Ivan Petrovi, Andrew Kellyii, Yulu Guoi, and Sarah La Monaca. 2019. "An Assessment of the Social Costs and Benefits of Vehicle Tax Reform in Ireland." Environment Working Papers, OECD, Paris.

Sargent & Lundy. 2009. "New Coal-Fired Power Plant Performance and Cost Estimates." Sargent & Lundy, Chicago, IL. https://www.epa.gov/sites/production/files/2015-08 /documents/coalperform.pdf.

Scovronick, Noah, Mark Budolfson, Francis Dennig, Frank Errickson, Marc Fleurbaey, Wei Peng, Robert H. Socolow, Dean Spears, and Fabian Wagner. 2019. "The Impact of Human Health Co-Benefits on Evaluations of Global Climate Policy." *Nature Communications* 10: 2095. https://doi.org/10.1038/s41467-019-09499-x.

Shabhazi, H., M. Reyhanian, V. Hosseini, and H. Afshin. 2016. "The Relative Contributions of Mobile Sources to Air Pollutant Emissions in Tehran, Iran: An Emission Inventory Approach." *Emission Control Science and Technology* 2 (1): 44–56.

Shindell, D. T., J. C. I. Kuylenstierna, E. Vignati, R. van Dingenen, M. Amann, Z. Klimont, S. C. Anenberg, et al. 2012. "Simultaneously Mitigating Near-Term Climate Change and Improving Human Health and Food Security." *Science* 335: 183–89. doi:10.1126/science.1210026.

Singh, D., S. Pachauri, and H. Zerriffi. 2017. "Environmental Payoffs of LPG Cooking in India." *Environmental Research Letters* 12 (11): 115003. https://iopscience.iop.org/article/10.1088/1748-9326/aa909d.

Slovic, A. D., M. A. de Oliveira, J. Biehl, and H. Ribeiro. 2016. "How Can Urban Policies Improve Air Quality and Help Mitigate Global Climate Change: A Systematic Mapping Review." *Journal of Urban Health* 93: 73–95. https://doi.org/10.1007/s11524-015-0007-8.

Smith, A. 2013. *The Climate Bonus: Co-benefits of Climate Policy.* New York: Routledge.

Sorrell, Steve, Birgitta Gatersleben, and Angela Druckman. 2020. "The Limits of Energy Sufficiency: A Review of the Evidence for Rebound Effects and Negative Spillovers from Behavioural Change." *Energy Research & Social Science* 64 (June).

Spencer, T., N. Berghmans, and O. Sartor. 2017. "Coal Transitions in China's Power Sector: A Plant-Level Assessment of Stranded Assets and Retirement Pathways." Study 11/17, Institute for Sustainable Development and International Relations, Paris.

Srivastava, Ravi K., Robert E. Hall, Sikander Khan, Kevin Culligan, and Bruce W. Lani. 2005. "Nitrogen Oxides Emission Control Options for Coal-Fired Electric Utility Boilers." *Journal of the Air and Waste Management Association* 55: (9): 1367–88. doi:10.1080/10473289.2005.10464736.

Srivastava, Ravi K., and W. Jozewicz. 2017. "Flue Gas Desulfurization: The State of the Art." *Journal of the Air and Waste Management Association* 51 (12): 1676–88. doi:10.1080/10473289.2001.10464387.

Taghvaee, Sina, Mohammad H. Sowlat, Amirhosein Mousavi, Mohammad Sadegh Hassanvand, Masud Yunesian, Kazem Naddafi, and Constantinos Sioutas. 2018. "Source Apportionment of Ambient $PM_{2.5}$ in Two Locations in Central Tehran Using the Positive Matrix Factorization (PMF) Model." *Science of the Total Environment* 628: 672–86.

Tibrewal, K., and C. Venkataraman. 2020. "Climate Co-Benefits of Air Quality and Clean Energy Policy in India." *Nature Sustainability* 4: 305–13. https://doi.org/10.1038/s41893-020-00666-3.

van Aardenne, John, Frank Dentener, Rita Van Dingenen, Greet Maenhout, Elina Marmer, Elisabetta Vignati, Peter Russ, Laszlo Szabo, and Frank Raes. 2010. "Climate and Air Quality Impacts of Combined Climate Change and Air Pollution Policy Scenarios." Working Paper 61281, Joint Research Centre, Seville. https://ideas.repec.org/p/ipt/iptwpa/jrc61281.html.

Vandyck, Toon, Kimon Keramidas, Alban Kitous, Joseph V. Spadaro, Rita Van Dingenen, Mike Holland, and Bert Savey. 2018. "Air Quality Co-Benefits for Human Health and Agriculture Counterbalance Costs to Meet Paris Agreement Pledges." *Nature Communications* 9: 4939. https://doi.org/10.1038/s41467-018-06885-9. https://www.nature.com/articles/s41467-018-06885-9.

Van Harmelen, T., J. Bakker, B. de Vries, D. P. van Vuuren, M. G. J. den Elzen, and P. Mayerhofen. 2002. "An Analysis of the Costs and Benefits of Joint Policies to Mitigate Climate Change and Regional Air Pollution in Europe." *Environmental Science and Policy* 5 (4): 349–65.

van Vuuren, D. P., and J. A. Bakkes. 1999. *GEO-2000 Alternative Policy Study for Europe and Central Asia.* Bilthoven: United Nations Environment Programme.

Vasquez Suarez, Claudia Ines, Feng Liu, and Grzegorz Peszko. 2018. *Scaling Up Thermal Retrofit of Residential and Public Buildings in Eastern Europe.* Washington, DC: World Bank. http://documents.worldbank.org/curated/en/572001543352247459/Scaling-up-thermal-retrofit-of-residential-and-public-buildings-in-Eastern-Europe.

Wang, L., P. L. Patel, S. Yu, B. Liu, J. Mcleod, L. E. Clarke, and W. Chen. 2016. "Win-Win Strategies to Promote Air Pollutant Control Policies and Non-Fossil Energy Target Regulation in China." *Applied Energy* 163: 244–53.

West, J. Jason, Steven J. Smith, Raquel A. Silva, Vaishali Naik, Yuqiang Zhang, Zachariah Adelman, Meridith M. Fry, et al. 2013. "Co-Benefits of Mitigating Global Greenhouse Gas Emissions for Future Air Quality and Human Health." *Nature Climate Change* 3: 885–89. doi:10.1038/nclimate2009.

WHO (World Health Organization). 2014. "Household Air Pollution and Health: Key Facts." WHO, Geneva. http://www.who.int/mediacentre/factsheets/fs292/en/.

Woodcock, James, Phil Edwards, Cathryn Tonne, Ben G. Armstrong, Olu Ashiru, David Banister, DeSean Beevers, et al. 2009. "Public Health Benefits of Strategies to Reduce Greenhouse-Gas Emissions: Urban Land Transport." *Health and Climate Change* 374 (9705). https://doi.org/10.1016/S0140-6736(09)61714-1.

World Bank. 2014. *FYR Macedonia Green Growth Country Assessment.* Washington, DC: World Bank. https://openknowledge.worldbank.org/handle/10986/19308.

World Bank. 2022. "Green Fiscal Reforms: Part Two of Strengthening Inclusion and Facilitating the Green Transition." Washington, DC, World Bank. https://openknowledge.worldbank.org/handle/10986/37308.

Wu, Zhangfa. 2001. *Air Pollution Control Costs for Coal-Fired Power Stations, CCC/53.* London: International Center for Sustainable Carbon. https://www.iea-coal.org/report/air-pollution-control-costs-for-coal-fired-power-stations-ccc-53/.

Xia, Ting, Monika Nitschke, Ying Zhang, Pushan Shah, Shona Crabb, and Alana Hansen. 2015. "Traffic-Related Air Pollution and Health Co-Benefits of Alternative Transport in Adelaide, South Australia." *Environment International* 74: 281–90. doi:10.1016/j.envint.2014.10.004.

Zlatev, Vasil Borislavov, Janusz Antoni Cofala, Grzegoz Peszko, and Qing Wang. 2021. "Clean Air and Cool Planet: Cost-Effective Air Quality Management in Kazakhstan and Its Impact on Greenhouse Gas Emissions." Washington, DC: World Bank Group. http://documents.worldbank.org/curated/en/099345012232191779/P1708700d2bd3a09093fa0cd27991d0662.

6 Integrated Policies on Air Pollution and Climate Change

INTRODUCTION

Policy makers do not implement abatement measures—polluting firms and households do, under the incentives created by policy makers. These incentives can be conveyed through cultural norms, economic incentives, and direct regulations, all influencing behavioral choices that decentralized economic actors make at all times. Policy makers have a duty to correct market failures discussed in chapter 2 and protect public health by designing policy incentives that encourage or force polluters to achieve targeted ambient air quality and reduce their carbon footprint. Policy makers are engaging in balancing acts between effectiveness and cost-efficiency of quickly reaching their air quality goals in the areas where the most people are exposed, while at the same facilitating structural transformation of the asset base to address longer-term climate change challenges. Sometimes air quality policies need to change to integrate climate-mitigation objectives. This chapter discusses how to soften tensions between the two agendas while leveraging those win-win opportunities that do exist. It also recognizes that the time scale needed to protect human health from air pollution is shorter than to achieve systemic decarbonization of energy, transport, and industrial systems.

Governments can use a wide menu of policy instruments to make polluters reduce the damage they inflict on the victims of pollution.

- *Direct regulations* prescribe or ban certain polluting technologies and activities, such as open burning of waste, driving diesel vehicles in city centers, or using certain fuels (for example, solid fuels or high-sulfur coal and fuel oil). Bans are as effective as the levels and enforcement of penalties for noncompliance. Effective enforcement requires strong governance and institutions (OECD 2009).
- Some direct regulations are implemented as *performance standards*, which leave emitters more flexibility on how to achieve required environmental performance. Examples include the US Corporate Average Fuel Economy (CAFE) standards and the emission-limit values under the EU Industrial Emissions Directive.

- *The most flexible instruments are economic in nature*, because they put a price on emissions, allowing emitters to freely choose how to react, and that reaction can range from paying the price and continuing to pollute to different ways of avoiding the price by reducing emissions.
- The softest, but essential instruments are *education, marketing campaigns, and behavioral nudges*. The latter draw on behavioral sciences and psychology to apply indirect suggestions as ways to influence the habitual behavior and choices made by groups or individuals (Thaler and Sunstein 2008).

Multiple policy instruments need to be applied jointly to tackle multiple environmental problems. Standard economic theory suggests that in the face of numerous externalities—local pollution and greenhouse gases (GHGs) in this case—the optimal policy is to use separate instruments for each, for example, a local pollution tax and a carbon tax rather than hoping that one instrument will solve other problems as a co-benefit (Hamilton et al. 2017; Pigou 1920). The general lessons learned from ex-post policy analysis and emerging empirical studies suggest that effective and efficient solutions to air pollution and climate policies require what the theory suggests—internalize both externalities with separate but integrated and coherent targeted policy instruments.

SYNERGIES AND TRADE-OFFS BETWEEN POLICIES ON AIR POLLUTION AND CLIMATE CHANGE

There is little evidence that climate policies implemented so far have had a significant impact on air pollution but strong evidence that improvements in air quality were attributed to targeted air pollution policies. The policies motivated by climate concerns played a minor, sometimes adverse, role in the strides toward better air quality (Andaloussi 2018; EEA 2019; Iyu et al. 2016; Li, Song, and Shen 2019; Massetti et al. 2017; Rafaj and Amann 2018; Rafaj, Amman, and Siri 2014; Rafaj et al. 2014; Singh 2017). Åström et al. (2017) apply decomposition analysis to determine the drivers that had the highest impact on SO_2 emissions during the period 1990–2012. They find that at least 26 percent of the decoupling of SO_2 emissions from economic growth was due to specific SO_2 policies (standards, permits, taxes, and fees), while the rest was driven by structural changes in the economy, productivity improvements in some industries, and fuel use changes, all weakly related to the historical stringency of climate policies.

Synergies and trade-offs between air pollution and climate policies depend on local conditions

Carbon prices and fuel taxes may reduce or increase air pollution depending on local conditions and policy design. The largest potential for synergies exists in the countries with a legacy of wasteful use of energy and the high share of coal in electricity generation, heating, and cooking. In such countries eliminating the most perverse fuel subsidies generates air pollution and climate co-benefits. For instance, Burtraw et al. (2003) simulate the effects of hypothetical relatively low carbon taxes for electricity production on reductions of NO_x emissions from US power utilities. They found potential for significant health-related ancillary air quality benefits through interaction between the carbon tax, stringent NO_x emission limit values applied in 1990, and the first large urban NO_x emissions trading program implemented in South Coast Air Quality Management District in 1994 (Burtraw and Szambelan 2009). Agee et al. (2014) find similar potential in their simulations of environmental policies applied to US power plants.

Econometric empirical studies are still few and localized, so it is premature to draw generalized conclusions from them. Nonetheless, Heger et al. (2019) show a statistically significant correlation between fuel prices and air pollution for urban transport in Greater Cairo (box 6.1). Kheiravar (2019) finds that each stage of fuel subsidy reform in the Islamic Republic of Iran brought diminishing improvements in ozone formation. Tan-Soo et al. (2019) used regression analysis to suggest that emissions of SO_2 and PM from industrial plants in Anhui (China) were inversely related to electricity prices, but this study contains some contradictions and did not cover air quality. While correlation between fuel prices and air pollution policies is quite well established in the literature, the evidence of causality is much weaker. Many variables usually determining the health impact of air pollution (such as targeted air quality management policies) are omitted in the regression models. Furthermore, econometric studies rarely explain the mechanism through which fuel prices could reduce air pollution. These transmission mechanisms are clearer in urban transport, where fuel prices are directly linked to quantity and type of fuel used, which in turn are more directly linked to

---

BOX 6.1

Empirical analysis of links between fossil fuel subsidies, public transport, and air pollution in Greater Cairo

Greater Cairo has a large and diverse vehicle fleet that accounts for almost one-third of particulate matter two-and-one-half microns or less in width ($PM_{2.5}$) concentrations. Most buses and trucks use diesel fuel running on old-generation diesel engines without catalytic converters or diesel particle filters. The government has adopted several policy measures over the past few decades, starting with the banning of lead in gasoline; improved car inspections; fiscal incentives for cleaner fuels (compressed natural gas) and for the use of newer cars, buses, and trucks; the retrofitting of fleets; and investments in alternative modes of transportation. The first two phases of the phasing out of fuel subsidies jointly reduced the number of cars driving in the streets of Cairo and reduced air pollution by 3.8 percent (the third wave of fuel subsidy removals has not yet been evaluated). Metro Line 3 (opened in 2012 and extended in 2014) reduced the number of cars driving in Cairo and reduced air pollution by 3.4 percent. Significant reductions in agricultural waste burning, in large part owing to the government's rice straw buy-back scheme, also contributed to cleaner air, though quantitative estimates of the impact are still pending. These interventions resulted in notable improvements in air quality. PM_{10} (particulate matter 10 microns in width or smaller) pollution decreased by nearly one-third from 2010 to 2017, but air pollution remains very high, such that Greater Cairo is still highly polluted. Using concentration-response relationships from the epidemiology literature alone shows that fuel subsidy removal and the opening of Metro Line 3 contributed to the avoidance of hundreds of infant deaths each year. Cost-benefit analysis reveals that this benefit alone is equivalent to about 10 percent of the metro's construction cost.

Source: Heger et al. 2019.

emissions and improved urban air quality. But for large stationary emission sources and residential heating and cooking using biomass, higher fuel and electricity prices may lead to unexpected behavioral and substitution effects and increase air pollution. Therefore, omitting these variables in econometric models can lead to misleading results. Overall, it is safe to say that empirical studies confirm hypotheses that policies increasing fuel prices are more likely to lead to improved air quality in the presence of significant energy subsidies, weak air quality management systems, and a legacy of a large and aged stock of energy- and emissions-intensive capital.

When the cheapest, win-win mitigation opportunities have been exploited, further increases of fuel prices confront the inelastic response of users/polluters. Excise duties on liquid fuels have been widely applied by the ministries of finance for a long time. The original motivation was not related to environmental benefits, but the relatively stable and progressive tax base. Taxes on transport fuels are convenient revenue raising instruments, because so far, fuel consumption was rising quite smoothly despite price fluctuations (the COVID-19 [coronavirus] pandemic was one exception to this rule because of lockdowns). But further increase of liquid fuel prices, extensions to other fuels and sources, and their carbon-related rate modulations lead to more complex interplay between incentives to reduce air pollution and incentives to mitigate climate change. The circumstances, under which this complexity can lead to trade-offs between climate mitigation and air quality agendas are illustrated with the following evidence:

- In transport, fuel tax differentiation motivated by, among other factors, fuel efficiency and climate concerns has historically promoted diesel vehicles. For example, in 2001, the UK introduced tax breaks for diesel cars to help the UK meet its obligations under the Kyoto Protocol. At that time diesel cars emitted an estimated 20 percent less CO_2 per kilometer than petrol cars. This surged their sales for years. Diesel support policies were reversed in Ireland (Ryani et al. [2019]), the UK, and the rest of the European Union (EU) only after 2015 when new information became available on their heavy toll on air pollution and after petrol engines made significant efficiency improvements. Overall higher fuel taxes have triggered innovation for more efficient internal combustion engines, more compact urban development (Bertaud 2003), and a modal switch to public transport, although the ultimate drivers that shaped all these trends were policy instruments and investments targeted at specific problems. Some anecdotal evidence suggests that with weak enforcement of targeted air pollution controls in transport, high fuel prices prompt owners of vehicles to have their diesel particle filters and catalytic converters removed to improve fuel efficiency and reduce operating and maintenance costs. Recent technology developments make the carbon footprint of diesel engines slightly greater than that of gasoline per kilometer driven, eliminating the climate rationale for diesel. The proposed revision of the EU Energy Taxation Directive marks a historical shift in the EU fiscal rules from favoring diesel to favoring petrol (European Commission 2021). In the shipping sector, low-sulfur fuel standards and sulfur taxes can have significant health benefits with a small climate penalty from reducing emissions of climate coolants. However, given that air pollutant emission standards raise fuel costs like carbon taxes, they also create incentives to enhance fuel efficiency in maritime transport and switch to low-carbon and low-pollution fuels such as natural gas and eventually hydrogen (Englert et al. 2021). Climate policies that favor biofuels (biodiesel, ammonia), however, could have a detrimental impact on air pollution in ports because of high primary particulate emissions and secondary $PM_{2.5}$ formation.

- Similar policy interactions can be observed in the stationary combustion sources. Climate policies encouraging the transition from fuel oil to biofuels in the heating sector reduce CO_2 emissions but often increase NO_x, PM, carbon monoxide, and volatile organic compound emissions (Brännlund and Kriström [2001]. An empirical analysis of Swedish heat and power plants (Bonilla, Coria, and Sterner 2018) finds that increasing carbon prices increased local pollutants (especially NO_x) in the absence of a commensurate steep increase in NO_x emission fees because the cost of CO_2 emissions (a sum of the carbon tax and Emissions Trading System [ETS] price) per unit of energy output during 2001–09 was higher than the cost of NO_x emissions. Ambec and Coria (2013) show that one critical factor determining whether increased stringency of climate policies leads to increased emissions of local pollutants is the elasticity of substitution between pollutants. If pollutants are substitutes, they argue, carbon prices will reduce carbon dioxide (CO_2) emissions and increase emissions of local pollutants. If they are complements, climate policies might lead to co-benefits because local pollutants will then be reduced alongside CO_2 emissions. They (as well as Bonilla, Coria, and Sterner 2018) suggest that in Sweden, climate and air pollutants "behaved" like substitutes in the period 2001–09 and that CO_2 emissions had higher abatement price elasticity than NO_x emissions, implying that carbon taxes unintendedly increased NO_x emissions. They also show that the effect of technological development can outweigh the substitution effect and decrease emissions of all pollutants. The authors conclude, "The fact that generating units face a trade-off between the pollutants indicates the need for policy coordination" (Bonilla, Coria, and Sterner 2018, 1). It should be stressed that in Sweden, an extraordinarily high carbon tax, sulfur tax, and NO_x emission fee were implemented almost simultaneously in the early 1990s and yielded significant reductions in emissions, quickly exhausting all available low-cost fuel efficiency and pollution abatement options. Most coal and heavy fuel oil power and heating plants had been converted to biofuels with stringent air pollution control equipment. In contrast, in the United States, carbon emissions trading and significant support for renewable energy were launched in the East Coast only in 2009, much later than NO_x emissions trading, and prices for both emissions were much lower than in Sweden leading to temporary trade-offs (see box 6.3). Going forward, Bonilla, Coria, and Sterner (2018) expect that further CO_2 emission reductions in Sweden would face harder trade-offs with NO_x emission intensities. However, Åström et al. (2017) show that in Sweden, which has already sharply reduced SO_2 emissions and simultaneously implemented stringent climate policies (the EU ETS and a national carbon tax), further tightening of air pollution policy instruments (standards, permits, fees, and the emission tax) would drive deep structural transformation and accelerated phase-out of remaining fossil fuels.
- In domestic heating and cooking, higher fossil fuel and carbon prices can encourage switching from natural gas, liquefied petroleum gas, or electricity to wood, other solid biomass or even waste, which has as large an adverse impact on human health as coal, especially given that biomass fuels are typically burned in small boilers or stoves where installation of end-of-pipe pollution control devices is not feasible (Bruce, Aunan, and Rehfuess 2017;

Cameron et al. 2016; European Commission 2021; IEA 2017; Kypridemos et al. 2020; Lee et al. 2013; Pachauri, Rao, and Cameron 2018; Peszko et al. 2019; Pittel and Rübbelke 2008; Republic of Serbia 2019).

- Sometimes higher fuel prices combined with weak air pollution policies have discouraged the application or effective operation of end-of-pipe pollution controls because of their energy penalty and operational cost. This occurred in the United States under the NO_x emissions trading program, when the NO_x allowance price fell below the cost of operating selective catalytic reduction (SCR) in coal power plants. The operators reacted by switching off their SCR systems and buying NO_x emission allowances instead. Air pollution increased for several years. The problem was rectified when the US Environmental Protection Agency reformed the system and tightened the NO_x emission caps. This increased the NO_x allowance prices, making operation of SCRs profitable again.

Only in the long term are stringent climate policies expected to eliminate most installations burning fossil fuels. In places with historically high fuel taxes (for example, Europe and Japan), the long-term rate of increase of fossil fuel demand was much slower than in places with low fuel taxes, such as the United States and net oil exporters (Sterner 2007). When fuel and carbon taxes are low, the least-cost way to improve air quality can be to install the state-of-the-art air pollution filters on the fuel combustion sources. Once air pollution controls are installed, increases in carbon prices can make it more profitable for plant operators to switch this equipment off if air pollution policies are not tightened at the same time. However, when air pollution policies are stringent enough, carbon and energy prices make the total costs of operating "clean" fossil fuel plants so high that operators may choose to retire the entire plant early leading to a more fundamental and structural decline of fuel-intensive economic activities. This process involves stranding existing assets and switching to technologies using cleaner energy sources (such as gas, which only emits NO_x) or nonbiomass renewables such as hydro, wind, and solar. The speed and depth of such a structural transition depends on many factors, including availability and costs of alternatives, whether competition is allowed in energy markets, and whether existing policies protect incumbents and preserve the status quo. Managing the transition requires massive infrastructure investments and behavioral changes to achieve scale. It also necessitates proactive social protection and labor market interventions to facilitate the transition without major social disruptions. Preparedness for such a transition varies greatly by country and is most challenging in countries dependent on fossil fuels (Peszko et al. 2020). Box 6.2 illustrates this complicated interplay between existing stringent air pollution policies and emerging climate policies in China.

Climate policies are more likely to have air-pollution co-benefits when strong targeted air pollution policies are in place

As a general principle, climate policies (for example, fuel taxes, carbon taxes, or emissions trading systems) should always include an assessment of their impact on air pollution. If a trade-off is likely, air quality regulations should be implemented or tightened at the same time. The stringency of both instruments (for instance, rates of carbon prices and air pollution prices) should be adjusted to account for the expected synergies and trade-offs until the time fossil fuels are phased out altogether.

BOX 6.2

Air pollution policies for coal power plants in China

Although air quality in many Chinese cities is still poor, significant improvements have been achieved. In the energy sector thousands of inefficient coal-burning units (small and large) were shut down and replaced with very efficient centralized heat and power plants, although often still fired by coal. A comparison done by the Center for American Progress shows that ultra-supercritical plants account for 92 percent of the top 100 most efficient coal-fired power units, while in the United States, fewer than 1 percent of the top 100 most efficient coal-fired power units are ultra-supercritical. This difference reflects the fact that the Chinese coal power plant fleet is much younger than in Organisation for Economic Co-operation and Development countries. Under the 11th Five-Year Plan (2006–10) the Chinese government began promoting the installation of flue-gas desulfurization units in thermal plants. By 2015, flue-gas desulfurization penetration reached 93 percent, which is higher than in the United States and even higher than in the European Union. Penetration of modern nitrogen oxide (NO_x) reduction equipment topped 50 percent. According to the National Aeronautics and Space Administration, sulfur dioxide (SO_2) emissions in China fell by 75 percent between 2008 and 2017 even though coal usage increased by approximately 50 percent and electricity generation grew by more than 100 percent in the same period. These results were achieved by a suite of specific air pollution policy instruments implemented by the government:

- The emission-performance standards for NO_x, SO_2, and particulate matter are stricter than comparable standards in the European Union and the United States (table B6.2.1). In highly polluted regions the permissible SO_2 emission standard for existing coal power plants is up to four times tighter than in the European Union and the United States. This high standard propelled operators to install expensive, high-efficiency postcombustion pollution control technologies and sometimes to switch to less-polluting fuels. In 2015, Beijing adopted the world's lowest emissions limit values for its coal power plants. A year later, it shut down four of them and switched to natural gas for heat and power generation (Singh 2017).

- The SO_2 emissions fee of 630 yuan per ton has been adjusted over time.

- Grants and subsidized loans are available to investing plants through various environmental funds to install air pollution controls.

- Since 2015, the penalties for noncompliance have been increased. The Environmental Protection Law introduced a cumulative penalty system, under which a noncompliant plant would be periodically penalized until it complies.

- The country has also introduced a preferential dispatch mechanism in certain areas, under which electricity generated by thermal power plants is scheduled on the central grid in accordance with the plants' air pollution rates and efficiency levels.

- Since 2004, coal-fired power plants with flue-gas desulfurization, and since 2011, plants with NO_x pollution-control facilities (selective catalytic reduction) enjoy a price premium for electricity they sell to the grid.

TABLE B6.2.1 Coal-fired power emission standards in China, the United States, and the European Union

Milligrams per cubic meter of exhaust gases

		CHINA	UNITED STATES	EUROPEAN UNION
Nitrogen oxides	Existing	100	135	200
	New	50	95	150
Sulfur dioxide	Existing	50/100/200[a]	185	200
	New	35	136	150
Particulate matter	Existing	20/30	19	20
	New	10	12	10

a. China's sulfur dioxide (SO_2) emission standards vary by location and age of plant. The strictest standards apply in key regions with the largest population exposure to air pollution.

Sources: Hart, Bassett, and Johnson 2017a; Hart, Bassett, and Johnson 2017b; NASA 2017.

The tightening of policies to control air pollution can have climate co-benefits when climate policies are in place. Slovic et al. (2016) review air pollution control policies and programs in megacities and find these policies can also contribute to mitigating and adapting to climate change. They observe that air pollution policies make operation of fossil fuel plants more expensive, reducing their commercial viability in competitive markets. The experience of coal power plants in Europe and the United States shows that important strides in the decarbonization of large power plants can be made by tightening air quality policies in the presence of stringent climate policy (or expectation of it). The US experience discussed in box 6.3 shows that expectations of high SO_2 allowance prices—which shot up to US$1,200 per ton in 2005—prompted massive investments in expensive flue-gas desulfurization (FGD) units, significantly increasing cost-recovery revenue requirements by coal power plants. The shale gas revolution and availability of cheaper (and subsidized) renewable energy sources rendered many such coal-power plants uncompetitive a few years later, leading to accelerated rates of their early retirement, especially since 2015. Extrapolating this experience indicates that the efforts to enhance competition in Chinese electricity markets discussed in boxes 6.2 and 6.6 and carbon pricing can exert strong competitive pressures on "clean" coal power plants bearing high fixed and operational costs associated with state-of-the-art air pollution control equipment.

Air pollution policies under some conditions can also harm the climate. Bans on biomass in household heating and cooking or financing its conversion to natural gas or coal-fired district heating reduce exposure to air pollution but increase emissions of CO_2 (although the ultimate impact on climate depends on case-by-case emissions of black carbon). Standards, taxes, or trading systems for SO_2 and NO_x emissions show a double climate penalty: First, they reduce emissions of climate coolants (sulfate and nitrate aerosols). Second, they induce installation of end-of-pipe filters that increase internal energy consumption by the combustion plant (see box 6.3 regarding the United States and box 6.2 regarding China). Therefore, application of carbon pricing in the absence of air pollution policies can lead to deteriorating air quality, while air pollution policies in the absence of carbon pricing can lead to costly carbon-intensive liabilities.

Figure 6.1 illustrates conditions under which climate and air pollution policies can reinforce each other, and conditions under which there may be trade-offs between them. The Integrated air quality and climate change approach is a balancing act between the following three policy conditions:

- Fuel prices include the full global social costs of burning those fuels.
- Air pollution policies internalize local costs of emissions and reduce population exposure to poor air quality in pollution hot spots.
- Air pollution *and* climate policies are applied jointly, although they are targeted separately at distinct externalities.

Policy design should always be tailored to local conditions and the nature of the air pollution challenge. Understanding the key local pollutants and sources responsible for poor air quality, least-cost abatement options, and political

FIGURE 6.1

Conditions under which air pollution and climate mitigation policies show co-benefits

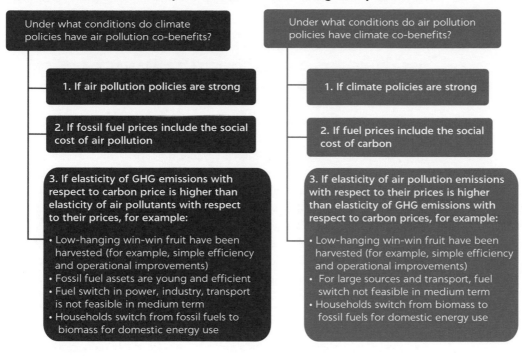

Source: World Bank.
Note: GHG = greenhouse gas.

economy and cultural and behavioral conditions helps dynamically adjust increasing stringency of air pollution policies to tightening of climate policies. Adequately tuned, integrated incentives encourage firms and households to take both climate and local pollution impacts into account in their operational and investment decisions. If these conditions are met, emissions of GHGs and air pollutants behave like complementary economic "bads," that is, an increase in the price of air pollution decreases emissions of climate pollutants and vice versa.

One example of recognizing the trade-offs as well as synergies between energy access, climate change, and air pollution agendas is the UN's Sustainable Energy for All energy access goal. Although reaching this goal is consistent with achieving a long-term 2°C warming target, it is recognized that it may result in some increase in global GHG emissions if liquefied petroleum gas is used on a large scale to achieve a significant portion of this access. This is considered a reasonable trade-off given the high social benefit of universal access to modern energy services (Rogelj, McCollum, and Riahi 2013).

Opportunities to integrate environmental and climate policies are real, but tensions between these two agendas are real as well. Misunderstanding these tensions leads to their disregard and hence mismanagement. Some trade-offs just need to be acknowledged and managed. Lack of awareness of these trade-offs can lead to the fallacy that implementing carbon taxes will solve the air pollution problem as a mere co-benefit and conversely that improving air quality will lead to the accelerated phasing out of fossil fuels. Such unfounded beliefs may underpin incoherent policy design and lead to unintended adverse impacts on air pollution or climate.

DESIGNING INTEGRATED POLICIES TO MANAGE AIR POLLUTION AND CLIMATE CHANGE

The integrated policy process proposed here focuses on the near-term health impacts of air pollution, while paving the way for long-term decarbonization. In the spirit of balancing multiple development objectives that are substitutes under some conditions and complements under others, this report proposes that the policies and measures that quickly and effectively prevent premature deaths and diseases from air pollution be prioritized so long as they do not create large irreversible climate liabilities in the future. The rationale for such a health-driven policy process is not universally accepted in public debate. Many authors, including some of those quoted in this volume, would argue that the policy process should always begin with policies promoting win-win abatement measures. The win-win (and co-benefit) narrative is based on the belief that air pollution and climate-mitigation objectives are always complements. As shown in this volume this is the case only if certain conditions are met, the most important being strong and enforceable air pollution policies that effectively reduce exposure to poor air quality in critical airsheds. If this condition is in place, health or win-win-driven policy processes should lead to similar outcomes.

Climate policies have their own strong rationale. Climate change is undoubtedly one of the main existential threats to humanity and the "greatest market failure the world has seen" (Stern 2008, 1). As climate change rises in the mainstream policy agenda, climate policies will no longer need to be justified by their local environmental co-benefits, especially because those benefits can be ambiguous. Climate policies can and should stand on their own and be implemented in coherent mixes with other economic and local environmental policies.

Climate policies are expected to phase out fossil fuels after 2050. However, fossil fuels are still included in the 2030 and 2050 energy mix in all International Panel for Climate Change (IPCC) 1.5°C and 2°C scenarios, as well as in the International Energy Agency scenarios (EPRI 2020). Therefore, as fossil fuels are phased out, air pollution policies are a necessary companion of climate policies, especially if fossil fuels are displaced by bioenergy, the use of which significantly increases in all 1.5°C and 2°C scenarios. In developing countries and countries with large numbers of people exposed to air pollution and limited resources to address multiple environmental challenges, the health-driven approach led by air pollution policies gains stronger political and social support since it prevents premature deaths and diseases each year.

One policy challenge is to integrate short- and long-term strategies. Transition to a low-carbon economy requires deep structural transformations, behavioral changes, and technological breakthroughs. The energy transition to renewable electricity, zero-emission vehicles, massive expansion of public transport, and zero carbon buildings have all gained political support and funding. Such projects offer the prospect of avoiding the compromises involved in making incremental improvements, but their costs as measured by investment requirements, delays, and general disruption tend to be challenging in the medium term, especially for developing countries. A large share of the reduction in emissions of air pollutants in the energy sector has been associated with a switch from coal to gas and installation of end-of-pipe pollution controls. The cost-effective and large-scale displacement of thermal sources affecting local air quality by nonpolluting renewable energy sources (such as wind, solar, hydro, or hydrogen) will take a long time and depends on factors that are beyond the control of local decision-makers in polluted communities (such as the decision to implement a carbon tax). Lvovsky et al. (2000) identify some rules of thumb for cost-effective strategies to address local air pollution while minimizing irreversible investments in carbon-intensive technologies and infrastructure. They differentiate between the time horizons in which the inherited capital stock of industrial, commercial, residential, and other assets can be altered.

- *The short-run* challenge is how to get the best performance out of existing assets. For example, simple modifications of the ways in which power or industrial burners are operated can lower NO_x emissions and improve their thermal efficiency. Using fuels with higher quality is another opportunity to improve environmental performance at low cost. Improvements in asset operation and maintenance can lead to higher levels of fuel efficiency (with climate co-benefits) and less local pollution per unit of fuel used. A combination of economic incentives, performance standards, sanctions, and behavioral nudges can be used to promote low-emission operation and improved management of existing assets to reduce emissions. These nudges could also involve gaining social acceptance for penalizing households for operating polluting vehicles and stoves, burning waste, or removing catalytic converters and diesel particle filters from diesel vehicles. In reality, the common approach to addressing political economy and social distributional challenges associated with improving the performance of existing assets is through simple derogations from higher performance standards without all the incentives and nudges proposed by Lvovsky et al. (2000).
- *In the medium term*, it is possible to shift the balance of the capital stock by ensuring that investment in new assets meets higher emission standards and is located where it does not add to high levels of pollution. In addition, new investments can be made to improve the environmental performance of existing assets through deep process improvements or end-of-pipe control technologies. The instruments that provide incentives for higher standards of rehabilitation investments can be implemented from the outset to enhance short-term benefits and avoid costly asset stranding in the future.
- *In the long run*, the capital stock may be regarded as largely or entirely flexible, and it becomes possible to achieve types of behavioral change that are not possible without large shifts in social norms, persistent habits, and long-lived physical assets. The time required to achieve fundamental long-term changes will vary across sectors but is rarely less than 15 years (for the vehicle fleet) and has traditionally exceeded 30–40 years for the energy and industrial sectors. For large combustion sources, structural transition away from fossil fuels requires heavy investment in new infrastructure that encompasses the span from generating and heating plants, on the one hand, to pipelines and transmission and distributions networks on the other hand. This transition may not happen quickly, especially in countries with mature infrastructure. However, in fast-growing developing countries where new investments are very large compared with the existing stock, such a transition can proceed

much faster. It is essential that all new investments in combustion plants as well as in system infrastructure be made in alignment with multiple local and planetary environmental boundaries—and that there be incentives for such alignments. Otherwise, the accumulation of new capital-intensive assets can lock in unsustainable development pathways that deplete natural capital upon which future prosperity depends. Multiple environmental crises or transition risks can turn these assets into liabilities that will be costly to pay for. For households, replacement of cooking and heating assets can be done quickly with modular investments. However, in the poorest communities in developing countries, many programs intended to provide access to clean fuels are notoriously difficult to scale up, especially when finance is limited for conversion to bottled natural gas or other "clean" fossil fuels.

In the absence of climate policies, air pollution policies can extend the life of capital- and carbon-intensive assets and thereby create potentially costly future liabilities. As discussed earlier, installation of end-of-pipe emission controls, such as FGD units, low-NO_x boilers or selective catalytic reduction systems, baghouses, or electrostatic precipitators at existing stationary coal or heavy fuel oil combustion plants is capital intensive and disruptive for operations. Therefore, such controls are usually piggybacked on major plant overhauls that require major investments and improve operating performance and thermal efficiency of the plant. Such investments can be induced by strict air pollution emission standards, permit requirements, or emission charges and taxes for conventional pollutants. Running pollution control equipment also involves additional costs of fuels, materials, and labor, further increasing the operating costs of "clean" plants when fuel and carbon prices are high. Before making large investments in air pollution controls, plant operators usually require assurance of a minimum operational life of the retrofitted plants to make a return on the capital invested (see box 6.3 regarding the United States). By providing such assurance through multiyear capacity payments, power-purchase agreements, or both, electricity-system operators assume contingent liabilities on behalf of consumers or taxpayers. These liabilities can become actual fiscal expenditures to keep these plants running or to pay stranded-asset costs when their operation is no longer legal or competitive. Therefore, introduction of air quality regulations, especially in the power sector, is a good time for power sector regulators to rethink long-term power system planning and establish a market design that facilitates a smooth and least-cost transition to efficient, clean, and low-carbon generating and transmission assets.

A coherent and integrated air pollution and climate policy framework would allow plant operators to optimize ongoing investments in capacity and future-oriented investments in innovation, jointly considering future air pollution and low-carbon policy incentives. In such coherent regulatory framework plant operators could make more informed choices between retrofitting existing coal-fired plants with state-of-the-art air monitoring and control equipment or switching to new, low-carbon technologies. Coherence between distinct climate and air pollution policy instruments would also ensure that some renewable energy sources, such as biomass and biofuels, do not create air pollution health hazards.

A health-led policy approach is particularly relevant for low- and middle-income countries. Such countries have contributed relatively little to the accumulated GHGs in the atmosphere and have limited fiscal space, as well as weak financial and institutional capacity, to address multiple local and global environmental issues at the same time. As demonstrated earlier, climate policies do too little too late to reduce the health risks of vulnerable citizens exposed to poor air quality if not worsening air pollution sometimes. Consequently, the sequencing and prioritizing of policies focused on environmental health looks rational and fair for such countries even if some of the policies may have a slight near-term warming impact on climate. Policies that encourage end-of-pipe pollution control technologies in existing plants or fuel switching from biomass to gas or smokeless briquettes for household cooking and heating can save many lives each year. In the most polluted regions, policies to improve local ambient air quality should stand on their own to deliver cost-effective results when and where needed most without having to prove climate co-benefits.

In higher-income and Organisation for Economic Co-operation and Development (OECD) countries, local air pollution policies can be more easily adjusted to enhance climate co-benefits without compromising air quality. This means prioritizing policies and measures that accelerate the structural transition away from fossil fuels over those that encourage retrofitting of existing assets that burn fossil fuels. However, even in the high-income countries of Europe, climate co-benefits are not a strict requirement for air pollution policies. Conversion of coal and biomass boilers to combined heat and power plants in district heating systems remains a legitimate and cost-effective measure for rapidly improving air quality despite a temporary negative impact on climate. Increasingly, advanced end-of-pipe emission-control equipment is required for a wide range of air pollutants from stationary and mobile sources despite their energy penalty and small increase in CO_2 emission intensity.

Instead of waiting for climate policies to solve air quality problems as a co-benefit, most OECD countries have implemented comprehensive packages of complementary regulatory instruments to target key sources of climate and air quality challenges. They have also put in place competition policies that facilitate rational long-term private investment and smooth the transition of capital to clean and low-carbon assets. For example, the local environmental footprint of large combustion sources operating in the European Union is regulated by the EU Industrial Emissions Directive, the National Emissions Reduction Commitments Directive, and the Air Quality Directive, while their global carbon footprint is controlled under the EU ETS and other climate policy regulations (for non-ETS sectors). Their coherent impact protects the health of the people now while facilitating long-term decarbonization across sectors and asset classes.

No single policy instrument can adequately address both air pollution and climate problems. Policy instruments that focus on mitigating climate change are not addressing pressing air pollution problems where and when needed. Their impact on climate is cumulative and measured in the time frame of decades whereas premature deaths and diseases caused by air pollution could be prevented instantly after emissions of air pollutants have been reduced. For example, a carbon tax should not be burdened with expectations that it will protect lives from air pollution as a co-benefit. Its job is to reduce GHG emissions. Carbon taxes are needed to drive long-term decarbonization even if they may temporarily deteriorate air quality by discouraging end-of-pipe air pollution controls or encouraging burning of biofuels. Rather than searching for the ambiguous air pollution co-benefits of carbon taxes, policy makers should from the outset design a coherent, dynamically adjusted package of climate and air pollution policy instruments. Such packages could include carbon taxes and local pollution taxes, together with direct regulations (for example, emission standards, fuel-quality standards for key local pollutants, or subsidies for households to meet their energy needs without air pollution). These taxes and regulations would be coupled with enforcement mechanisms, behavioral nudges, and capacity strengthening. Such an integrated design avoids shifting the problem from one dimension to another, saves money, and mitigates social confusion. It also mitigates the risk of conflicting incentives and future policy reversals.

Policy design for air pollution, unlike for climate mitigation, must consider local geography, topography, atmospheric chemistry, and meteorological conditions. The relevant impacts of climate policies are global. The location of sources of CO_2 and other GHGs that are uniformly mixed in the atmosphere does not matter for their warming potential, except for black carbon, the radiative forcing of which is short lived and localized. In contrast, policy design for air quality should always begin with identification of pollutants and sources that are the main contributors to the exposure of populations in the specific airsheds with poor air quality. As discussed earlier, the key harmful pollutants in the ambient air, such as $PM_{2.5}$ and ground-level ozone, are often not emitted directly but are formed in the atmosphere from emissions of precursor pollutants (though a fraction of $PM_{2.5}$ concentration also comes from "primary" particles). Therefore, policy interventions need to target those precursors. Some sources of precursor emissions are far away from exposed people and are not visible to the naked eye.

Location of emissions is critical for air quality and irrelevant for climate change. Strasert, Teh, and Cohan (2019) analyze the climate benefits and air quality benefits of closing coal-fired plants in Texas and find that although CO_2 emission reduction rates are fairly similar across plants, the local health benefits are highly plant specific. The plants' impacts on ozone, $PM_{2.5}$ formation, and associated health and visibility outcomes vary by an order of magnitude depending on where the plant is located and whether it has been fitted with pollution control equipment. Therefore, many countries differentiate the rates of emission taxes or standards for emissions of precursor air pollutants by geography (see the discussion of China in box 6.2, the United States in box 6.3, and Chile in box 6.7). Also, bans on the most polluting fuels and equipment are often limited to locations where they cause dangerous health impacts to large populations living in the most polluted airsheds. Source-apportionment studies and emission dispersion photochemical models help identify transmission channels between emission sources and population exposure to air pollution.

Policy design should consider the seasonality of air pollution. At different times of the year, policy instruments and their enforcement need to focus on different pollutants and pollution sources. For example, open burning of agricultural waste is highly seasonal and happens after harvest. In the Northern Hemisphere the main air pollution problem is winter smog caused by high concentrations of particulates, SO_2, and aromatic hydrocarbons related to heating sources. This was the case with the 1952 Great Smog of London and current pollution hot spots in Eastern Europe; Ulaanbaatar, Mongolia; China (Zhao et al. 2019); and the Western Balkans (see figure 6.2). In contrast, ground-level ozone (related mainly to transport pollution) creates summer smog in cities with warmer climates (famously first experienced by Los Angeles) because ozone formation requires a lot of sunlight. Though policy efforts need to consider seasonality of pollution sources, an excessively narrow focus on seasonal controls,

FIGURE 6.2

Source attributions to modeled PM$_{2.5}$ pollution over an annual cycle in two selected Western Balkan cities (monthly average, 2018)

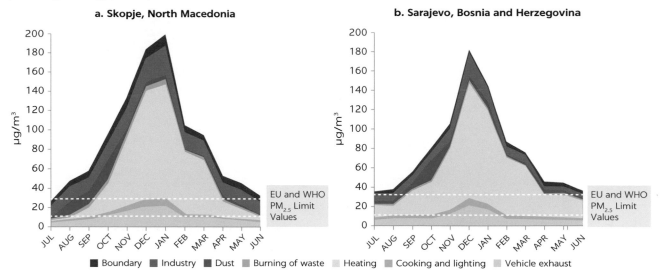

Source: Modeling executed by UrbanEmissions.info for World Bank, November 2019.
Note: Units on vertical axes µg/m³; EU = European Union; PM$_{2.5}$ = particulate matter two-and-one-half microns or less in width; WHO = World Health Organization; boundary = emissions coming from outside of the city borders.

especially in the large stationary combustion sources (power and industry), can lead to inefficient investment choices and costly adjustments later, as with NO$_X$ emissions trading for US power plants (see box 6.3).

The US (and EU) examples show that environmental policy is a dynamic process of designing instruments with the best current knowledge and then adjusting them to surprises. Some of these adjustments are needed to manage emerging trade-offs and synergies between local and global pollution. Figure 6.3 illustrates the evolution of the NO$_X$ and SO$_2$ programs in the United States and the dominant pollution abatement measures used by regulated emission sources in response to evolving package of clean air policy instruments. In the period 1995–2018, all subsequent NO$_X$ and SO$_2$ emissions trading programs in the United States led to significant reductions of both pollutants. During the same period, CO$_2$ emissions from the same power plants increased or remained stable until 2010, following trends in electricity production. Thus, reduction of local pollutants from thermal power plants did not lead to climate benefits in this period. In fact, Andaloussi (2018) notes that installations of FGD units coincided with small increases in CO$_2$ emissions. CO$_2$ emissions declined only when cheaper gas and renewable-energy power plants began to displace coal power plants, stranding some of them before they reached the end of their expected economic lifetimes.

The main sources of air pollution often change over time; consequently, policy instruments must be dynamically adjusted while not deviating significantly from long-term decarbonization pathways. For example, probably the world's first national air quality law—the United Kingdom's Clean Air Act, enacted in 1956, principally in response to London's Great Smog of 1952, and amended in 1968, focused on policies to control domestic coal burning in stationary sources, with some climate co-benefits, mainly from switching

Consistency and flexibility in the air quality and climate policy process in the United States

NO_x trading programs

The NO_x Budget Program was a cap-and-trade program created in 2003 to reduce nitrogen oxides (NO_x) emissions in the eastern United States during the warm summer months, when NO_x interacts with sunlight to create the highest concentrations of ground-level ozone, causing serious respiratory diseases in the densely populated East Coast. By 2008, the NO_x Budget Program dramatically reduced NO_x emissions from targeted power plants and large industrial sources during the summer months (US EPA 2009). Because only summer emissions were capped, most coal power plants invested in selective catalytic reduction (SCR) systems, which were relatively cheap to install but expensive to operate, and switched them off between September and May when no allowances were needed to offset emissions (McNevin 2016).

In 2009, the NO_x Budget Program was replaced by the Clean Air Interstate Rule (CAIR), and the US Environmental Protection Agency introduced emission caps on annual NO_x emissions that were also binding in the winter (US EPA 2010). Emitting plants could trade allowances across all 28 states in the Eastern United States. NO_x seasonal and annual budgets were met by the plants covered by CAIR, driving the price of NO_x allowances below the marginal cost of operating SCR units. Meanwhile, the shale gas revolution, as well as rapid deployment of low-marginal-cost wind and solar power plants, both triggered by the stimulus programs following the 2008 financial crisis, made gas, wind, and solar power generation unexpectedly competitive. Under the competitive pressure from low-carbon power generators and with low prices for NO_x emission allowances, several coal plant operators found it cheaper to switch off the SCR units and buy low-cost allowances, leading to increases in NO_x emissions from the SCR-equipped plants (McNevin 2016) and an increase in ozone concentrations during the period 2008–12 (Carlson and Burtraw 2019).

The government reacted by replacing the CAIR in 2015 with the Cross-State Air Pollution Rule. Caps were tightened, interstate trading was no longer allowed, and prices and incentives to use SCR recovered.

SO_2 trading programs

The Clean Air Act of 1970 established national standards for ambient air quality for sulfur dioxide (SO_2). Power utilities have cost-effectively reduced their contributions to local air pollution by building tall stacks and dispersing their emissions high into the atmosphere (Regens and Rycroft 1988). This solved the local problem but allowed the particles to travel for hundreds of kilometers to precipitate as acid rain in down-wind states, causing significant damage (Schmalensee and Stavins 2012).

To rectify this problem, in 1995 an Acid Rain Program established a market-based cap-and-trade system for SO_2 emissions from all US power plants. The program delivered emissions reduction faster and at lower cost than expected. Plant operators in the eastern United States devised a cost-effective compliance strategy by switching from local high-sulfur coal to the low-sulfur coal shipped from the western United States. US railways were deregulated in the late 1970s, reducing the costs of freight transport and making this option cheaper than installing flue-gas desulfurization units (scrubbers). According to the US Energy Information Administration, fuel switching accounted for 59 percent of emissions reduction and scrubbers were installed at only about 10 percent of the units, accounting for 28 percent of emissions reduction.

New information about the role of SO_2 in the creation of hazardous secondary particulate matter two-and-one-half microns or less in width ($PM_{2.5}$) led to the CAIR program in 2005, aimed at reducing SO_2 emissions by 70 percent below the 2003 level. Amid expectations about tightening emission caps and the possibility of banking allowances from the previous program, the prices of SO_2 allowances skyrocketed, reaching US\$1,200 per ton in 2005, justifying massive investments in expensive flue-gas desulfurization units. By 2008, overinvestment in flue-gas

continued

Box 6.3, *continued*

desulfurization and unexpected regulatory uncertainty amid legal challenges from industry depressed allowance prices to trivial levels after 2010, shaving off financial returns on the earlier pollution control investments.

The Cross-State Air Pollution Rule program introduced in 2015 tightened the cap. The court disallowed interstate trading of allowances, limiting access to cheap abatement opportunities outside state boundaries. Allowance prices increased but remained too low to cover the fixed and variable costs of flue-gas desulfurization units. Competitive pressure from cheaper natural gas and renewable energy sources rendered many "clean" coal-fired power plants uncompetitive, leading to accelerated rates of their retirement after 2015.

FIGURE 6.3

Evolution of US NO$_x$ and SO$_2$ emissions trading programs for thermal power plants

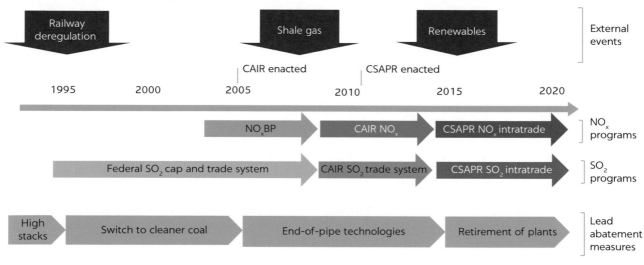

Source: World Bank, based on various US Environmental Protection Agency sources.
Note: The bottom arrows show the sequencing of the dominant SO$_2$ and NO$_x$ abatement measures used by the power plants' operators during the evolving emissions trading programs in the United States. The figure illustrates that the combined impacts of regulatory incentives, external events, and expectations encouraged abatement responses in the order of increasing marginal costs, eventually leading to the most expensive abatement option of retiring coal power plants, the only measure with strong climate co-benefits. BP = Budget Program; CAIR = Clean Air Interstate Rule; CSAPR = Cross-State Air Pollution Rule; NO$_x$ = nitrogen oxides; SO$_2$ = sulfur dioxide.

from coal to natural gas for heating and power. Beginning with its 1993 amendment,[1] the policy focus began shifting toward emissions from vehicles. Box 6.4 illustrates the dynamic process of air pollution and climate policy development with dynamic adjustment to new emerging challenges and lessons learned about policy results and interactions.

More recently, agricultural emissions, which contribute to the formation of secondary PM$_{2.5}$, have been recognized as an important source of air pollution in cities. For example, air quality improvement efforts in Paris have historically focused on transport and diesel engines, but attention is now broadening to include regulating the sources of ammonia, methane, and NO$_x$ from agriculture (Petetin et al. 2016). Similar features are observed in the most polluted cities in developing countries (Air Quality Expert Group 2018;

Evolution of air quality management and climate policies in Mexico

In Mexico City air quality management began in the 1990s with the regulatory requirement for the state-owned oil company to phase out lead from gasoline sold in the country. This regulation was aimed at addressing health issues, with no climate co-benefits. In the following years, a suite of measures was gradually introduced to reduce other air pollutants from city traffic, such as restrictions on the use of passenger cars on certain days of the week, the requirement for catalytic converters and particulate filters for passenger cars, and the tightening of vehicle pollution standards backed by improved inspection and enforcement. Eventually Mexico embarked on infrastructure investments in public transport—the long-term, capital-intensive program with the largest climate co-benefits. While the most toxic heavy metals and most visible particulate pollution was significantly improved and the passenger car fleet became cleaner, invisible ozone concentrations, invisible fine particles, and heavy-duty vehicles were targeted by a second wave of regulations. It was not until 2013 that Mexico introduced an economywide carbon tax on fuels and 2020 that the pilot emissions trading system was launched (World Bank 2009, 2020b).

Fuller 2018). New policies enacted at the national or EU level also target agricultural sources of urban air pollution. For example, the EU National Emissions Reduction Commitments Directive restricts the use of ammonium carbonate fertilizers and calls on member states to find ways to reduce ammonia emissions from inorganic fertilizers and other sources, such as manure management. Most recently, Hebei province in China targeted reduction of the excessive use of nitrogen fertilizer after finding that it accounted for one-third of $PM_{2.5}$ pollution in Beijing (World Bank 2020b). Air quality in many urban areas in Australia, California, Mexico, and the Russian Federation is becoming increasingly affected by seasonal forest or bush fires, which are likely to increase in frequency and intensity along with global warming. The evolution of policy design for the control of SO_2 and NO_X emissions from large combustion sources in the United States (discussed in box 6.3) shows that flexible design and the ability to adjust the policy package as new information becomes available encourages cost-effective abatement responses and strengthens integration of agendas on air pollution and climate mitigation.

PRICES OR QUANTITIES: THE CHOICE OF AIR POLLUTION POLICY INSTRUMENTS

Economic and fiscal instruments offer certainty about the ceiling for abatement costs (which cannot be higher than a tax rate). However, for precisely that reason, the environmental effect is harder to predict, because it is difficult to predict how polluters will react to price signals. Prices give polluters free choice about whether to abate emissions or continue polluting and pay the price for it. The policy rule of thumb is that when the health costs of pollution are more uncertain and potentially high, the policy mix should, as a precaution, rely on quantity-based instruments, which deliver higher effectiveness when enforced (Baumol and Oates 1971, 1988; Weitzman 1974). Therefore, toxic pollutants (such as heavy metals, dioxins, aromatic hydrocarbons, and so on) that create significant health hazards for large, exposed populations should be controlled mainly through quantitative restrictions, whereas price-based policy instruments are more appropriate for less locally harmful and more mixed pollutants, such as GHGs. Most OECD countries regulate emissions of hazardous air pollutants, such as mercury, from power plants mostly through performance standards, emission limit values, or requirements for meeting certain technology performance specifications (Maximum Achievable Control Technology for the US Environmental Protection Agency or Best Available Techniques in the European Union).[2] The best candidate for price-based policy instruments is CO_2 because it is perfectly mixed in the atmosphere, it does not matter where the emission source is located, and the local harm is negligible. Charges or taxes[3] are usually applied on some air pollutants, such as SO_2, NO_x, and PM, as a complementary incentive on top of source-specific quantitative emission standards. Emissions trading systems combine the precautionary nature of quantitative instruments (because total emissions from a group of sources is capped) with the efficiency of price instruments, because participants can trade emission allowances between themselves, achieving the least-cost allocation of emissions between sources without government intervention. The United States was the first country to introduce large-scale emissions trading systems for SO_2 and NO_x (box 6.3), though in 2019 Gujarat state in India piloted the first emissions trading system for PM (https://gpcb.gujarat.gov.in/webcontroller/page/emissions-trading-scheme-pilot-project).

Both emission standards and taxes or charges can and often should be applied to the same pollutant emitted from the same source category. Countries such as France, Poland, Sweden, and the United States apply pollution charges to SO_2 and NO_x (and PM in Poland) emitted from combustion sources that also must comply with emissions limit values and comprehensive performance standards under the EU Industrial Emissions Directive or equivalent US emission standards (Åström et al. 2017; Bergquist et al. 2013). With this policy package, the emission limit values put a static cap on emission factors, while emission taxes or charges dynamically encourage plant operators to reduce emission intensity even below the standard. Direct regulations can also be designed to induce dynamic innovation. For example, the Best Available Techniques under the EU Industrial Emissions Directive include emission limit values for all installations in a sector that are established through public-private dialogue inspired by the best performers in the industry. Every few years the regulator reviews the environmental technology and management innovation in an industry and tightens up emission limit values to match the progress made by the best performers. In this way firms have incentives to be at the cutting edge of innovation and have their performance standards imposed on competitors. (See box 6.5 for a comparison of Poland's and the United Kingdom's experiences.)

A policy instrument will be effective and socially acceptable only if affected firms and households have access to affordable alternatives. Availability and affordability of alternative energy sources and their carbon footprint determine the level and direction of climate co-benefits. For example, small back-up electricity generators fueled by diesel and petrol became a significant source of fine particulate pollution in Lagos, Nigeria, because of the unreliability of grid electricity (Croitoru, Chang, and Kelly 2020). Power sector reform initiated in Nigeria is expected to increase the reliability of grid electricity while at the

BOX 6.5

Regulating installations, fuels, or both? The experiences of Poland and the United Kingdom

Controlling emissions from small combustion sources such as household heating and cooking stoves is complicated. These stoves can accommodate a wide range of solid fuels, and monitoring what they really burn is difficult. Whether the same stove burns coke, briquettes, wood, or even domestic waste makes a big difference for air pollution. The UK Clean Air Acts of 1956 and 1968 aimed to control air pollution by introducing smoke control areas (now covering most urban centers) where the burning of wood and coal is not permitted, but exemptions are granted for the burning of "authorized fuels" or the "exempt appliances" listed and updated on the government website "Smoke Control Areas: The Rules" (https://www.gov.uk/smoke-control-area-rules).

For many years, municipalities in southern Poland tried to reduce winter smog by encouraging the conversion of small heating sources to modern solutions (mainly gas and district heating, but also modern solid fuel stoves) and preventing the burning of the most-polluting fuels, especially coal and household waste. However, checking what households burn—even in modern solid fuel stoves—was difficult. Environmental inspectors had to catch people in the act of burning low-quality hard coal waste, lignite, or even household trash (Dworakowska et al. 2018). In recent years drone technology with chemical sensors has been used to catch perpetrators, but monitoring and enforcement remained expensive and cumbersome. Therefore, in September 2019, Krakow became the first Polish city to ban the burning of any solid fuels in any boilers or stoves with a thermal capacity of less than 1 megawatt (larger installations are regulated with environmental permits). Coal producers objected to the ban, preferring to regulate equipment performance standards rather than broad equipment types or fuel types, and to limit restrictions to low-quality coals and stoves not meeting the high standards of the EU Ecodesign Directive.

The resistance eventually withered away, especially after the European Commission restricted EU structural funds from being used to finance conversion to any installations burning coal. The ban was widely supported by the population of Krakow because of the vibrant clean air social movement during the previous 30 years, the ban's strong analytical underpinning, and almost three decades of public financial support for conversion of individual solid-fuel boilers to cleaner alternatives. Other cities in southern Poland followed Krakow's example and began phasing in restrictions on the use of solid fuels. Furthermore, the Krakow example prompted the draft amendment to the National Energy Policy presented by the Ministry of Climate in September 2020, which sets a target for the total phaseout of the use of coal in domestic heating by 2030 in urban area and by 2040 in rural areas. This process is an example of how multiple layers of direct and economic regulation can not only strengthen the enforcement of air quality improvement efforts, but also improve alignment between air pollution and climate policy goals.

same time reducing air pollution and delivering climate benefits because grid power supply is dominated by gas-fired plants. Morocco maintains subsidies for the domestic use of butane for heating and cooking at a very high fiscal cost amid concerns that this is the only available alternative preventing households from returning to the use of wood and charcoal, which would increase air pollution, deforestation, and desertification. There is some discussion about redirecting some of the subsidies to boost domestic production of solar water heaters, which are currently imported and unaffordable by low-income households, but developing domestic manufacturing takes time and would also not solve the cooking problem (Peszko et al. 2019). Klausbruckner et al. (2016)

show that pricing policies implemented to mitigate climate change might have increased negative health effects because of an unanticipated increase in local air pollution in South Africa, and these indirect consequences must be taken into account when devising mitigation strategies. Addis Ababa, the capital of Ethiopia, introduced the electric light railway as an affordable alternative to polluting minibuses that had dominated city public transport. Its success enhanced mobility for low-income people and demonstrated significant synergy between air pollution and climate change given that most of the country's electricity is generated by hydropower plants.

Market and fuel-pricing reforms have a significant impact on the emission of air and climate pollutants. The market-driven collapse of heavy, inefficient industries in Eastern Europe after the fall of the Soviet Union has been the major driver of improved air pollution and carbon intensity alike, although commercialization and privatization of polluting enterprises also helped enforce environmental regulations. More recently, the targeted air pollution regulations for coal power plants in China, supported by power procurement rules, led to the world's lowest SO_2, NO_x, and PM emission intensities, but may have created an expensive capacity and carbon bubble in the power sector, which can complicate planned electricity market reform, accelerated penetration of renewables, and introduction of carbon pricing—all essential to implementing China's commitment to becoming carbon neutral by 2060 (see box 6.6).

BOX 6.6

Air pollution, climate policies, and the power market design impact on coal power plants in China

Until recently, operators of coal-fired power plants in China obtained a fixed regulated "benchmark offtake price" from grid operators. Their operating hours, however, have been declining, undermining plant efficiency. In some provinces the grid operators call on the coal power plants to produce electricity only after first dispatching renewables, nuclear, and natural gas power plants. Furthermore, the dispatch sequence between coal power plants has been determined by their thermal efficiency (heat rate), and then by generators' sulfur dioxide (SO_2) emissions. In addition, operators of the plants equipped with state-of-the-art SO_2 and nitrogen oxides control equipment have been granted a preferential tariff premium for electricity sold to the grid. These pricing and dispatch preferences for cleaner plants, alongside other air pollution control policies discussed in box 6.2, resulted in the largest, and one of the most efficient and "locally cleanest" fleets of coal power plants in the world.

However, continued building of new coal power plants despite slowing electricity demand and increasing penetration of cleaner gas and renewables resulted in excess installed coal capacity except in a few provinces where availability of renewables is limited. To allow revenues to cover the costs of operation and debt service, coal power plants have obtained guaranteed utilization hours.

Against this background, China launched a gradual power-market reform aimed at lowering power costs to consumers.

- In 2015, the State Council issued Decree No. 9, which started a gradual reduction of the guaranteed operating hours for coal power plants and an increase of the share of bilateral contracts between buyers and sellers. The operating hours of coal power plants declined below 50 percent on average, squeezing the profit

continued

Box 6.6, *continued*

margins of the operators. In some provinces, such as Shanxi, a dispatch policy no longer guarantees any quantity or price for new coal power and gives priority dispatch to renewables.

• On January 1, 2020, China replaced its fixed benchmark on-grid tariff for coal generation with a more flexible mechanism with a "floating" tariff component determined through negotiations between plant operators and electricity buyers.

These reforms of the power market design, together with stringent emission standards for local pollutants and with excess capacity, have further increased competitive pressure on coal-power generators, which are already struggling to be profitable with low-capacity utilization rates. A significant increase in investments in new coal-fired generation would add to a capacity glut and further squeeze operating hours and profits of coal power plants. In addition, the recent Chinese pledge of carbon neutrality by 2060 could result in more aggressive actions to reduce coal consumption and further reduce coal capacity utilization and significantly increase grid penetration by renewables, leading to premature retirement of the least efficient and most polluting units, with significant benefits for both air pollution and climate, but with major legal, financial, social, and political economy challenges.

The power market reforms and efforts toward carbon neutrality will be a stress test for China's stringent pollution control regulations. As discussed above, most grid-connected coal power plants in China are new, efficient, and fully equipped with expensive air pollution control facilities. Consequently, coal plants that remain in the grid may be under financial pressure to use cheaper, poorer-quality coal; further restrict operating hours; and importantly, turn off pollution control equipment to reduce operating costs to stay competitive with less carbon-intensive gas or renewable plants. Operation at partial load conditions can have an adverse impact on plants' thermal efficiency as well as on air pollution and greenhouse gas emission intensity.

This evolving Chinese experience and the US experience described in box 6.3 are humbling lessons, suggesting that no regulatory and institutional system can be designed perfectly from the outset and stay unchanged forever. The alignment between electricity markets, climate and air pollution polices, and other social, economic, and environmental objectives is an ongoing process of gradual adjustment of multiple instruments to respond to external changes and increase policy coherence over time.

Sources: Burnard et al. 2014; Hao 2020; Hart, Bassett, and Johnson 2017a, 2017b; Myllyvirta, Zhang, and Shen 2020; Singh 2017; Thompson 2020; Yan and Wu 2017. Also Xiaodong Wang (World Bank), personal communication, October 2020.

Contract structure in the energy sector can lead to unwanted environmental effects. In many developing countries coal power plants are built cheap, without advanced air pollution controls and under long-term (20–30 year) power purchase agreements that guarantee minimum operating hours and offtake prices and contain "take-or-pay" clauses. For example, only 1 percent of the total coal-fired power plant capacity in India has FGD units (IEA 2020). Under the prevailing commercial contract structures, national regulators find it challenging to demand that these plants be retrofitted with pollution controls or to retire the noncompliant plants early without compensation. Renegotiating the power purchase contracts is possible in principle, but the resulting increase in offtake prices would need to be shared between the utility, electricity consumers, and plant operators. Examples in China and South Africa show that it can be done, and the legal, political economy, and social distributional challenges can be addressed. The government of India is preparing to face these challenges through the coal sector commercialization and electricity market reforms launched in 2020.

Policy design needs to be realistic about administrative capacity. Developing countries may not have the resources and institutional capacity to design and enforce complex policy packages to facilitate fast systemic transformation involving massive, accelerated asset turnover and behavioral change. In countries with weaker institutional capacity, policies need to be designed for easy enforcement. Even OECD countries often face challenges enforcing low-emission behavior, especially from small, decentralized sources, such as vehicles and homes. As discussed earlier, several EU countries reported problems with illegal removal of diesel particulate filters from vehicles. Mexico City designed enforceable rules to control vehicle pollution by prohibiting the use of vehicles older than six years in the city. In 2015, however, Mexico's Supreme Court ruled that all vehicles that passed smoke inspection tests, even older ones, must be allowed to operate in town. This triggered a reversal of smog frequency because drivers found it easy to bribe the workers at inspection stations to get polluting cars approved.[4] The government responded by modifying vehicle inspection and maintenance protocols and extending the regulations to government vehicles and other heavy-duty vehicles to address this and other enforcement challenges (International Transport Forum 2017).

FISCAL INSTRUMENTS TO ADDRESS AIR POLLUTION AND CLIMATE CHANGE

Green fiscal instruments send a price signal to reduce emissions but leave firms and households with flexibility about how to modify polluting behavior. Environmental taxes set a price on emissions, either downstream per unit of a pollutant emitted or upstream by adjusting fuel or other product taxes according to the pollutant content. This price signal makes the environmental costs of using fuels visible to firms and households, hence prompting adjustment of their investment and behavioral choices. These "technology-neutral" price signals are, above all, promising efficiency. There are two dimensions of efficiency. *Static efficiency* reduces the total cost of emission reduction for all regulated economic agents by allocating emission abatement efforts to those with the lowest cost until marginal abatement costs across firms are equal. *Dynamic efficiency* prompts economic agents to continuously innovate and find new, low-cost ways to reduce emissions to avoid paying charges on the marginal unit of emissions. Taxes or charges on air pollutants come across as a convenient policy instrument for developing countries with multiple preexisting fiscal and environmental policy distortions. Environmental taxation is often recommended as part of a broader environmental regulatory and fiscal reform, especially if a country has weak institutions, relatively few large polluters operating in noncompetitive markets, and a large informal sector.

The current architecture of energy taxation often includes perverse environmental incentives, encouraging both global and local pollution. OECD (2019) found that coal and heavy fuel oil, the most carbon-intensive and the most air polluting fuels, were (in 2018) taxed at the lowest rates of all fuels or are not taxed at all (see figure 6.4). Energy taxes on diesel are lower than on gasoline in all countries studied by OECD (2019) except Mexico, Turkey, and the United States (where fuel taxes are much lower than in the rest of the OECD). Energy taxes on natural gas are higher than they are on coal.

FIGURE 6.4

Effective energy tax rates and their carbon benchmarks for 44 OECD and selected partner economies and for international transport

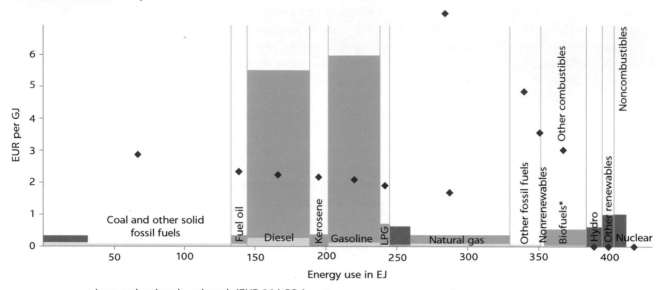

Source: OECD 2019.

Note: Weighted average by energy category and energy end use (electricity or other). Tax rates applicable on July 1, 2018. The energy use is for 2016 and adapted from IEA 2018 *World Energy Statistics and Balances*. The energy base does not include electricity and heating imports, for which the primary energy source is not known. Biofuels are marked with an asterisk since carbon dioxide emissions from the combustion of biofuels are considered zero in the greenhouse gas inventories reported under the United Nations Framework Convention on Climate Change. EJ = exajoule; LPG = liquefied petroleum gas; OECD = Organisation for Economic Co-operation and Development.

Many countries provide indirect and implicit subsidies to domestic fuel use by waiving excise taxes on coal, oil, and gas used by industry (Peszko et al. 2019). In Western Balkan countries, fuel taxes are often below the minimum rates of the EU 2003 Energy Taxation Directive (ETD) and misaligned with the social costs of fuel use (see table 6.1). For example, coal and natural gas used for heating purposes in North Macedonia and Serbia are not taxed at all. Interestingly, diesel in the EU ETD was taxed at a lower rate per liter than petrol, which translates to even larger tax favor for diesel fuel per unit of energy, because of the higher energy density of diesel versus petrol. Among the selected countries listed in table 6.1, only Serbia taxed diesel correctly, that is, not only above the minimum EU levels, but also above the tax rate for gasoline, reflecting its much worse impact on air pollution and health. The legislative proposal for the revised EU Energy Taxation Directive submitted by the European Commission in 2021 significantly increases the minimum tax rates and links them more closely to energy content and environmental impact. Taxation of biomass and biofuels is also introduced, differentiating between sustainably and unsustainably sourced fuels. This revision proposed to tax diesel fuel for transport at the same rate as gasoline, €10.57 per gigajoule, which implies a higher tax rate per liter (€0.41/liter for diesel versus €0.37/liter for gasoline). However, the explicit carbon and air pollution components of the energy excise taxes considered by the European Commission staff (European Commission 2021) did not find its way into the final legislative proposal, at least so far.

TABLE 6.1 **Excise duty rates in selected Western Balkan and EU countries**

	BULGARIA	MONTENEGRO	NORTH MACEDONIA	SERBIA	SLOVENIA	2003 EU DIRECTIVE	UNITS	2021 EU PROPOSAL FOR 2023 RATES
Unleaded petrol	0.363	0.549	0.353	0.488	0.547	0.36	€/liter	€10.75/GJ
Diesel as a motor fuel	0.33	0.44	0.246	0.502	0.469	0.33	€/liter	€10.75/GJ
Diesel heating (nonbusiness)	0.33	0.207	0.1	0	0.234	0.02	€/liter	€0.9/GJ
Natural gas (motor fuel)	0.43	0	0	0	3.74	2.6	€/GJ	€7.17/GJ[a]
Natural gas heating (business)	0.31	0	0	0	1.85	0.15	€/GJ	€0.6/GJ[b]
Natural gas (nonbusiness)	0	0	0	0	1.85	0.3	€/GJ	€0.6/GJ[b]
Coal and coke (business)	0.31	0.3	0	0	2.34	0.15	€/GJ	€0.9/GJ
Coal and coke (nonbusiness)	0.31	0.3	0	0	2.34	0.3	€/GJ	€0.9/GJ

Source: Based on World Bank (2020a) and European Commission (2021).
Note: EU = European Union; GJ = gigajoule.
a. Tax rate for natural gas as motor fuel to reach the same rate as petrol by 2033.
b. Tax rate for natural gas as heating fuel to reach the same rate as coal by 2033

Several countries levy charges or taxes on air pollutants, with SO_2, NO_x, and PM being the most common emission tax bases (see table 6.2). Chile (box 6.7), France, Poland (box 6.8), and several Eastern European countries levy taxes or charges downstream on emissions measured or estimated in the flue gases at the end of the stack.

Pigouvian air pollution taxes can be calibrated to the external damage caused by pollution, to the marginal abatement cost of the desired emission reduction, or to revenue targets. According to classic literature (Baumol and Oates 1988; Pigou 1920), pollution tax rates should be linked to marginal damage costs and therefore should differ not only by pollutant but also by their location and impact on the exposed population. Simulations using a model of the California electricity sector show that a location-based tax on air pollutants can double the health benefits related to air quality compared with a carbon tax with similar GHG reductions (Weber 2021). Only Chile had previously attempted to design such a close-to-optimal environmental tax, in 2017 (see box 6.7). The Chilean nationwide differentiation of air pollution taxes by airshed is an experiment that involves high transaction costs and requires strong environmental management institutions. An initial assessment of environmental outcomes looks promising, but longer-term impacts still need to be studied rigorously to offer lessons learned for other countries. Nordic countries calibrate their NO_x and SO_2 tax rates or fees to the estimated marginal cost of targeted emission reduction (Svenningsen et al. 2019), which may be similar to the marginal cost of damages done by pollution. Most countries calibrate air pollution tax rates or fees to revenue targets and differentiate rates by the relative average environmental impact, subject to political and social feasibility. Several countries, such as Australia, Canada, and Spain, allow subnational entities (provinces) to determine their own pollution tax rates (OECD 2021).

TABLE 6.2 **Selected environmental taxes and charges**

	POLLUTANT TARGETED	TAX RATE AND BASE	COVERAGE	USE OF REVENUES
Sweden SO_2 tax	SO_2	€3,300/ton of sulfur content of fuel	All fuels entering economy (sulfur content > 0.05%)	General budget
Sweden nitrogen charge (feebate)	NO_2	€5,500/ton	Stationary combustion plants > 25 MWh of useful energy per year	Rebated to industry in proportion to energy sales (after administrative cost)
France SO_2 tax	SO_2	€140/ton	Large combustion sources	General budget
Poland SO_2 and NO_x charges	SO_2 and NO_x	€130 ($143)/ton	Combustion sources > 5MW	Earmarked to off-budgetary environmental fund
Poland PM charges	PM (soot) particles	€357 ($398)/ton	Combustion sources > 5MW	Earmarked to off-budgetary environmental fund
Chile	PM, NO_x, SO_x	$500–$60,000/ton depending on location and damage	Large combustion sources > 50MWt	Not available

Source: National statistical offices.

Note: MW = megawatt; MWh = megawatt hours; MWt = megawatts thermal; NO_2 = nitrogen dioxide; NO_x = nitrogen oxides; PM = particulate matter; SO_2 = sulfur dioxide; SO_x = sulfur oxides.

Direct regulations (if enforced) offer certainty that individual sources and the industry as a whole are not exceeding critical emission thresholds, whereas emission charges provide dynamic incentives to continuously seek innovative and cheaper ways of reducing emissions. In Sweden, for example, NO_x feebates were levied on top of existing emission standards and created dynamic incentives to reduce emissions and increase innovation and efficiency of power plants at the same time. The Swedish environmental taxes coexist not only with multiple direct regulations under the EU directives but also with energy tax preferences, a renewable electricity certificate system, and private-public research and development initiatives. Jointly this mix of instruments has led to the massive displacement of oil and coal by sustainable biomass in district heating systems. The comprehensive, integrated, and well-designed regulatory framework induced innovation breakthroughs in solid and liquid biofuel technologies and in efficient biomass combustion technologies, not only in district heating but also in industry (Swedish Forest Agency 2018; Wei et al. 2013). Polish and French air pollution charges coexist with a suite of direct regulations under the EU directives, such as the Air Quality Directive and the Industrial Emissions Directive, that require all sources to meet strict emissions-limit values for the same pollutants and to apply the Best Available Techniques to control air pollution. These quasi-fiscal instruments and direct regulations applied to the same sources are incentive-compatible and complement each other. Emissions prices also make switching off filters uneconomic when fuel or carbon prices increase (see box 6.2). Box 6.8 reviews selected examples of air pollution tax design in Europe.

The combination of environmental taxes, expenditure policies, and complementary policies became known as environmental fiscal reform (by the OECD) or environmental tax reforms (by the International Monetary Fund [IMF]) (see figure 6.5). Economists have long debated whether and under what conditions Environmental Fiscal Reform can yield economic and fiscal benefits even before the direct welfare benefits of avoided pollution are counted (the so-called double dividend). The available analysis of the fiscal dividend

BOX 6.7

Downstream air pollution taxes in Chile

Chile faces severe environmental problems related to both the impacts of climate change and local atmospheric pollution. To help address these challenges and contribute to climate mitigation efforts, the government of President Bachelet designed a two-pronged approach implementing environmental taxes that target local and global pollution simultaneously. In a General Tax Reform Bill (Law No. 20.780) passed in September 2014, two pollution taxes were introduced as of 2017, affecting large stationary sources with boilers or turbines (with an aggregate capacity of 50 megawatts or more).

The taxes target technologies and facilities across different economic sectors such as food processing, refineries, and electricity. They include a small flat carbon tax and a series of local emission taxes (on particulate matter [PM], nitrogen oxides [NO_x], and sulfur dioxide [SO_2]), the rates of which vary depending on location. The PM tax rate can vary from US\$500 to as much as US\$60,000 per ton. The rate on carbon dioxide (CO_2) emissions was set at US\$5/ton. The tax on local pollution (PM, NO_x, and SO_2) is set at a variable rate depending on the estimated environmental damage that a marginal unit of emissions generates in a specific locality or municipality. To capture the environmental damage, the legislation set a per capita rate for each contaminant, and calculates the tax rate based on a formula that depends on the per capita rate times the number of inhabitants in a local municipality, and a coefficient for carrying capacity (tax per ton of pollutant i in municipality j = 0.1 × carrying capacity (1, 1.1, or 1.2) × per capita tax of pollutant i × population of municipality j). For example, the tax rate, in the case of PM, can vary from US\$500 to US\$60,000 per ton, depending on the air quality in the affected airshed and number of exposed people.

Although further research is necessary to evaluate the full impact, the taxes have proved to be extremely effective. In 2017, 96 facilities were responsible for raising US\$191 million in tax revenues. The CO_2 tax covered approximately 40 percent of the country's carbon emissions. All electric generation plants publicly declared that they would not implement future coal-based installations. Moreover, energy companies signed an agreement with the Energy Ministry to dismantle existing coal plants.[a] For local contaminants, evidence indicates that facilities have introduced abatement equipment that has considerably reduced their emissions. Although these investments are motivated by companies' efforts to reduce their tax burden, they have had a significant environmental impact.

These taxes can and should be improved over time. The tax bases can be broadened, rates can be raised, or both. A compensation mechanism or offsetting scheme can be designed for the CO_2 tax. However, what is relevant is that Chile has shown a way forward for developing and middle-income countries to tackle their most significant environmental problems—local and global pollution—in a consistent and coherent manner with market-based economic instruments.

Source: Contributed by Rodrigo Pizarro, former head of the Division of Environmental Economics, Ministry of Environment of Chile.
a. GOBIERNO Y GENERADORAS ANUNCIAN FIN DE NUEVOS DESARROLLOS DE PLANTAS A CARBÓN (http://generadoras.cl/media/page -files/391/180129%20Comunicado%20no%20mas%20nuevas%20plantas%20a%20carb%C3%B3n%20-%20ME%20MMA%20 Generadoras%20de%20Chile.pdf).

from environmental taxes has so far focused on fuel and carbon taxes. Bovenberg and van der Ploeg (1994), Bovenberg and Goulder (1996), Bovenberg and de Mooij (1997), Goulder (1995), and Mooji (1999) argue that when complex tax interactions are considered, the fiscal dividend is questionable, but in recent literature the World Bank (Pigato 2019) and the IMF (IMF 2019) agree that carbon taxes, at least, can deliver double dividends under the condition of preexisting distortions in the economy, which are quite common, especially in developing countries. In imperfect world growth can be enhanced by taxing labor, income, and capital less and energy and pollution more. Shifting the tax

BOX 6.8

Examples of air pollution tax designs in Europe

NO_x feebate in Sweden

In Sweden, nitrogen oxide (NO_x) fees are levied on thermal power and heating plants and large combustion sources in industry per unit of their NO_x emissions. Firms pay in proportion to their total annual NO_x emissions. The rate was set at the level of the marginal cost of pollution abatement to reach the Swedish policy target. At €5,500 per ton of NO_x it is the highest fee on NO_x pollution in the world (although local NO_x tax rates in Chile can in principle be higher). Revenues are rebated back to paying firms in proportion to their electricity sold to the grid, after deductions for administrative costs for the system (therefore they are sometimes called feebates). The government forgoes the revenue, so such feebates are technically not fiscal instruments, not taxes. In this way some plants make a net profit while the rest make a net payment within the system. Firms emitting low volumes of NO_x per unit of energy produced are net beneficiaries of the scheme. The feebate design creates incentives for each plant to reduce emissions and at the same time to increase generation efficiency so as to be dispatched more often to the grid. Because the fee revenues are recycled within the sector, the negative impact on international competitiveness is eliminated and the operators of the most efficient plants have access to resources to reduce their NO_x emission intensity. The incentive effect of the feebate was enabled by the competitive design of the Nordic electricity-only market (Nordpool), which is the most competitive electricity-only spot market in the world. In 1992, the first year the tax was applied, total revenues were 612 million Swedish kroner). In 2011, total tax revenues amounted to 794 million Swedish kroner even though NO_x emissions per unit of energy produced fell by more than 50 percent. This revenue increase was possible because the tax base increased over time as smaller firms became subject to the tax (OECD 2013).

Upstream sulfur tax in Sweden

The Swedish sulfur tax is levied on the sulfur content in multiple fuels upstream when the fuel is sold to combustion plants. Fuels used for manufacturing lime, stone, and cement and in soda boilers in the pulp and paper industry are wholly exempt from the tax. Diesel and heating oils used for shipping, trams, and railways, in addition to aviation fuel, are exempt from the tax. A portion of the tax revenues received by the government is reimbursed to the payers in proportion to the sulfur that was removed or fixed in ashes and therefore not emitted to the atmosphere. By collecting the tax first and reimbursing later, the government shifts the burden of abatement proof onto taxpayers. The administration of pollution taxes requires interagency coordination. The tax administration relies on the environmental administration to check the polluters' emission abatement reports, which increases transaction costs. However, under the EU Industrial Emissions Directive, all large combustion plants already must have continuous emissions monitoring systems in place.

Air pollution charges and environmental protection funds in Poland

The system of Polish environmental charges and funds dates to the late 1980s. Charges are levied on several hundred air and water pollutants and on waste materials deposited in landfills. More than a hundred air pollutant and source categories are subject to paying for emissions. The legal title of the charges is "fees for the use of environmental services," and they are introduced not through the tax code but through environmental protection laws. The rates are indexed every year and enacted through a resolution of the Council of Ministers. The rates are high by international standards except Scandinavia and Chile (see table 6.2), although their main goal is to raise revenue for environmental expenditures. The base for these charges is extremely wide, covering 67 air pollutants plus special charges for air emissions from small combustion sources (less than 5 megawatts thermal), internal combustion engines, and reloading of liquid fuels. There are also special charges for air pollution from poultry farms, where rates can vary by a factor of 10 depending on the type of farm, type of birds, and farming practices. In 2019, total revenue from environment-related charges and fees amounted to PLN971 million (about US$262 million), of which air

continued

Box 6.8, *continued*

pollution charges alone delivered about US$122 million. However, this revenue has been declining for several years mainly because of falling emissions and the introduction of rate ceilings. All revenues are earmarked for environmental funds controlled by different levels of government. Polish fees for the use of environmental services are quasi-fiscal instruments. Revenues from environmental charges and the earnings and expenditures of environmental funds are reported in annexes to regional and state budgets.

Source: Åström et al. 2017; OECD 2013; Peszko 1999; SEPA and Swedish Energy Agency 2007; Statistics Poland 2020; Sterner and Isaksson 2006.

FIGURE 6.5

Environmental fiscal reform and its potential double dividend

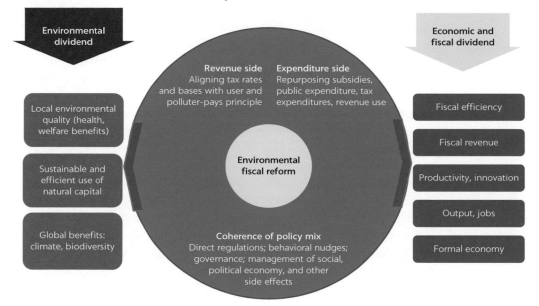

Source: Based on original analysis for this publication.

burden from economic "goods" to economic "bads" and taxing rents rather than profits can broaden the tax base and improve fiscal efficiency while reducing tax distortions on entrepreneurship, innovation, and growth.

Carbon and fuel taxes are usually better revenue raisers than air pollution taxes. Improvement in fiscal efficiency through environmental fiscal reform is more likely to result from upstream taxes or charges on uniformly mixed pollutants, such as CO_2, or just fuel excise taxes. Carbon and fuel taxes can be easier to administer, easier to collect, and more difficult to evade than direct taxes and even value added taxes (Liu 2013). As de facto excise taxes, they are a convenient fiscal revenue source in countries with weak fiscal institutions and a large informal sector (Pigato 2019). Carbon and fuel taxes can be imposed and collected upstream in the fuel value chain using existing excise tax administration. Their incentive impact on GHG emissions is always positive because the only way to reduce CO_2 emissions from fuel combustion is to reduce fuel use. The amount of carbon embedded in the fuel is directly linked to, and always the same as, the carbon emitted to the atmosphere during its combustion because there are no commercially available technologies to remove CO_2 from flue gases. Therefore, the carbon tax base can be simply defined as carbon content per ton of fuel used (or per energy content of fuel). Upstream tax design also does not compromise environmental effectiveness because the location of GHG emission sources does not matter for the value of damage.

The revenue from fuel and carbon taxes is predictable and relatively stable. Finance ministers want tax revenues to be predictable over time, and the tax base to be either stable or, even better, grow and decline automatically with economic cycles (as value added taxes and income taxes do). The base of carbon and fuel taxes does not erode quickly given that price demand elasticities for most fuels are low in the short and medium term and the income effect counters the price effect. GHG postcombustion abatement technologies, such as carbon capture and storage, are not proven on a commercial scale, so the only way to reduce GHG emissions is to use less fuel or displace the equipment that uses it. Decarbonization requires massive upfront investments to replace existing produced capital stock, massive migration of labor, and fundamental changes involving behavior and culture. Few cost-competitive substitutes for fossil fuels are yet available outside of the power sector. Therefore, demand for fuels (especially oil and its derivatives) is relatively inelastic with respect to price in the short and medium term. Low price elasticity of demand is one reason that transport fuels are a preferred base for excise taxes in most countries (OECD 2021). Another reason is that consumption of fuels, especially transport fuels, has been growing with income, outstripping demand reduction due to price increase. Such taxes act like automatic stabilizers and allow governments to increase budget revenue without periodically going through the politically difficult process of raising tax rates to finance increasing expenditure needs.

The revenue base for pollution taxes is expected to erode more quickly than that for fuel and carbon taxes. This is an intended effect because air pollutants cause larger, more immediate, and greater local health damage than GHGs. Air pollution taxes are expected to cut emissions of conventional pollutants much faster than carbon taxes are expected to cut the use of fossil fuels. Furthermore, reducing emissions of local pollutants is often easier than reducing emissions of GHGs. Broad availability of affordable end-of-pipe pollution control technologies allows significant improvement of air quality without reducing fuel use and requiring major structural transformation. The more a pollution tax is designed to reduce emissions rather than raise revenue, the more difficult it is to plan a balanced budget with it. Figure 6.6 illustrates that in the period 1993–2008, the revenues from the sulfur tax in Sweden eroded dramatically with the drop in emissions and remained negligible compared with carbon tax revenues, as a result of successful mitigation. In contrast, carbon tax revenues initially increased before stabilizing.

Pollution taxes can still play an important role in tax policy. Ministries of finance have widely accepted taxation of "sin products," such as tobacco and alcohol, where reduction of their consumption, and thus erosion of the tax base, is expected among the tax policy objectives. Revenues from pollution taxes can be stabilized either by increasing the rates of existing taxes and fees or by broadening the tax base to additional air pollutants, water pollutants, chemicals, nonrecyclable consumer goods, virgin plastic resins, and so on. For example, in Sweden revenues from the bundle of various pollution taxes and fees steeply increased in 2018 as a result of the introduction of a new tax on polluting chemicals (illustrated in figure 6.6).

The more effective a pollution tax is in reducing population exposure to air pollution, the more complex tax administration tends to be. In principle, air pollution taxes can also be imposed upstream as part of fuel excises, making tax administration easier. However, doing so may compromise their environmental effectiveness depending on circumstances. First, the upstream design does not differentiate between emission sources, even though the population exposure and value of damage by pollution is highly dependent on where emissions occur and on the profiles of individual emission sources. For example, low-stack sources located upwind of fragile airsheds with common atmospheric inversions, like the Indo-Gangetic plains or several cities in the Western Balkans, inflict several times more harm to health than emission sources located downwind of exposed populations or in well-ventilated areas, such as along the seacoasts. Second, the amount of polluting substance contained in the fuel is directly proportional to the damage caused by pollution only for sulfur (as discussed above). For the most common air pollutants, such as $PM_{2.5}$ and tropospheric ozone, a large portion of the ambient concentration originates from secondary formation of these particles in the atmosphere from emissions of precursor gases. Tropospheric ozone is entirely formed in the atmosphere from precursor gases (methane, volatile organic compounds, NO_x, and carbon monoxide) in the presence of sunlight. Typically, between 30 percent and 50 percent of $PM_{2.5}$ in ambient air consists of secondary particles formed from precursor emissions, such as SO_2, NO_x, volatile organic compounds, and ammonia. According to recent research, these secondary particles, especially those originating from sulfates, are more harmful to health than primary $PM_{2.5}$ particles (World Bank 2021). Therefore, upstream taxes on the $PM_{2.5}$ content of fuels are not directly related to the external damage caused by individual sources, giving rise to suboptimal abatement choices. Similarly, NO_x emissions from combustion sources partly

FIGURE 6.6

Revenues from environmental taxes in Sweden, 1993–2018

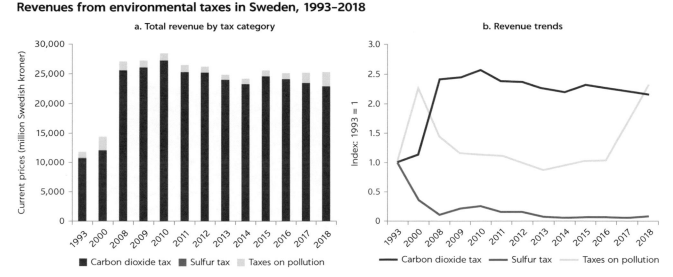

Source: Based on data from Statistikmyndigheten SCB (https://www.scb.se/en/finding-statistics/statistics-by-subject-area/environment /environmental-accounts-and-sustainable-development/system-of-environmental-and-economic-accounts/pong/tables-and-graphs /environmental-taxes/total-environmental-taxes-in-sweden/#Fotnoter).

come from the nitrogen content in the fuel (fuel NO_x) and partly from the nitrogen contained in the air (thermal NO_x and prompt NO_x). The share of fuel NO_x is higher for combustion of coal than gas, and the share of thermal NO_x can be high or low depending on the combustion conditions (mainly combustion temperature). Third, unlike GHGs, air pollutants can be removed before emissions using multiple cost-competitive abatement measures. This challenge can, in principle, be rectified by upstream taxes and downstream rebates, but this design proved complicated for air pollution taxes.

The design and implementation of pollution taxes cannot be done by tax administration alone. Even in countries where environmental taxes are well integrated into fiscal policy, such as SO_2 taxes in France and Sweden and pollution taxes in Chile, fiscal administration relies on technical support from the line ministries responsible for environment, usually for design, rate setting, monitoring, and verification. The Indian state of Gujarat implemented one of the world's first emissions trading systems for particulate matter (Greenstone et al. 2019). The upstream design was considered poorly related to emissions and difficult to implement fairly. Therefore, the price was levied based on independently certified continuous emissions monitoring of total suspended particles as a proxy for health damage. Continuous monitoring of $PM_{2.5}$ was not feasible. Measuring emissions at the end of the stack is a more environmentally effective way of determining the tax base than taxing fuels used as the inputs to combustion. Some countries (Poland, for example) determine pollution fees due for smaller sources based on emission factors that differ by the specific source characteristics.

Of the conventional pollutants, SO_2 is the best candidate for upstream design because the sulfur content in fuel is easy to measure and proportional to end-of-pipe emissions without any controls. Upstream taxes with downstream rebates, in principle, can help address the principal-agent problem and reduce the costs of tax collection by pushing the burden of proof of abatement on polluters when they claim refunds. The challenge is that for all conventional pollutants (unlike for CO_2) there are plenty of process-oriented or end-of-pipe technologies available to remove pollutants from fuel before combustion or from flue gases after combustion. In principle, a portion of tax revenue can be refunded to polluters upon proof of how much sulfur was removed before emitting to the atmosphere. So far, only the Nordic countries (Sweden and, in narrower scope, Denmark and Norway) collect SO_2 taxes upstream at fuel-distribution choke points—with the tax base being the sulfur content of the fuel. Tax authorities refund a portion of tax revenues to payers downstream, after evidence of removal of sulfur from fuels before emission has been verified. This design has not been replicated elsewhere, however, because of legal issues and concerns about transaction costs and liquidity risks for firms. Poor governance and lack of trust in the government increases firms' concerns that refunds may be used for rent extraction by authorities. For large and medium combustion sources, improved technology for continuous emissions monitoring systems dramatically reduced transaction costs and firms' information advantage worldwide, often making direct taxation of emissions administratively easier than managing downstream refunds. However, this design is still worth considering, among other options, as an economic instrument for reducing SO_2 emissions, especially in countries with traditions of weak enforcement of environmental regulations and large information asymmetries between polluters and environmental administration. Upstream taxation is an efficient way of internalizing the external costs of products that are polluting during or after the consumption phase (such as plastics or harmful substances). Another example of the successful application of upstream emissions pricing was a trading system for the quotas of lead that could be added to gasoline in the United States in the 1980s (Ellerman, Joskow, and Harrison 2003; Newell and Rogers 2003).

Separate taxes on air pollution and carbon emissions should be applied jointly to mitigate the risk of aggravating one environmental problem while solving another. As discussed extensively in this report, climate policies, such as carbon prices, can increase emissions of air pollutants by discouraging the installation and use of air pollution control equipment unless strong air pollution regulatory controls (and taxes) are in place. Climate policies can also encourage substitution from "clean" fossil fuels (natural gas, liquefied petroleum gas, or petrol) to more polluting bioenergy. Power plant and vehicle operators react differently to energy price increases in the presence than in the absence of air pollution regulations and taxes. For example, in Sweden CO_2 and energy taxes have helped high sulfur taxes and direct regulations reduce both SO_2 and CO_2 emissions by shifting consumption from coal and oil to biofuels. As discussed earlier, climate-motivated energy policies have led to a temporary worsening of air pollution in the United States, amid low NO_X emission prices. There is widespread, though only anecdotal, evidence that households in Central and Eastern Europe have been switching from gas and district heating to coal and biomass and even household waste for residential heating in individual stoves, when gas and thermal electricity became expensive. Similar concerns are raised for the power sector in Brazil (Portugal-Pereira et al. 2018). To mitigate such risk, Morocco is maintaining fossil fuel (butane) subsidies despite a high fiscal burden because authorities are concerned that phasing them out would push poor rural households back to deforestation and the burning of biomass and waste in household stoves (Peszko 2019). Some evidence also suggests that switching off filters at night in stationary plants or removing catalytic converters from vehicles is common in several countries that either do not have, or do not enforce, air pollution policies.

In the opposite direction, as also discussed above, air pollution policies in the absence of climate policies can increase global warming through at least four different impact channels: (1) the energy penalty associated with equipment for air pollution control, (2) lower emissions of climate coolants such as SO_2 and NO_x, (3) switching from bio-energy to natural gas, and (4) extension of the economic lifetime of carbon-intensive assets fitted or retrofitted with costly pollution control equipment. The first three channels usually lead to only slight, temporary warming and just need to be accepted as a small price to pay for saving lives and protecting public health from air pollution. The scale of potential warming declines as new pollution control technologies become more and more efficient in energy use. Installation of pollution controls is usually accompanied by the broader overhaul of the whole plant. Therefore, improved energy efficiency of retrofitted installations often outweighs the energy penalty added by the pollution control equipment. The large-scale risk of locking in carbon-intensive infrastructure can be managed by the policy incentives for new and retrofitting investments in fossil fuel assets that jointly optimizes for the costs of climate and air pollution mitigation. The economic incentives are coherent when carbon prices are jointly levied on the same installations with explicit prices on SO_2, NO_x, PM, and other relevant air pollutants to reflect the comprehensive external costs to society.

CONCLUSION: TOWARD POLICY COHERENCE AND POLICY INTEGRATION IN APPROACHES TO AIR POLLUTION AND CLIMATE CHANGE

The integrated air quality and climate change (IAQCC) policy process requires that targeted climate and air quality regulations be implemented jointly, calibrated to harness win-win opportunities when relevant at the airshed level, and mitigate the risk of aggravating one environmental problem while solving another. The IAQCC is a dynamic process of learning, reviewing, and adjusting policy responses to evolving air quality challenges and climate change challenges. This process focuses on the near-term health impacts of air pollution in affected airsheds while paving the way for long-term global decarbonization. Its five key steps are illustrated in figure 6.7.

An integrated approach to developing and implementing an air quality and climate change policy process has its foundation in the two pillars of policy coherence and policy integration and consists of five steps:

The first step is to establish a ground-level air quality monitoring system to determine where indoor air pollution and ambient air pollution (mainly tropospheric ozone and $PM_{2.5}$) reach concentrations that pose a high potential risk to health.

The second step is to (1) determine the need for action by examining the exposure of populations, assets, and ecosystems to poor air quality in these

FIGURE 6.7

Five steps of integrated air quality and climate change (IAQCC) policy process

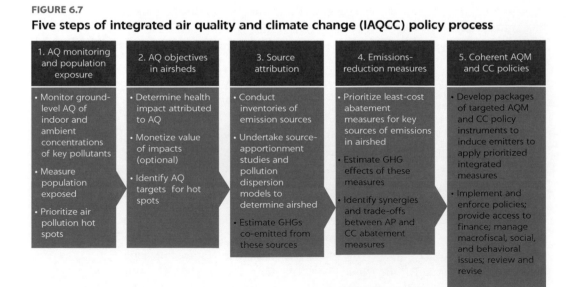

Source: World Bank staff.
Note: white text denotes actions driven primarily by health considerations. Black text refers to actions that gradually introduce an integrated approach to mitigating air pollution and climate change. AP = air pollution; AQ = air quality; AQM = air quality management; CC = climate change; GHGs = greenhouse gases.

hot spots and (2) estimate the actual impact that such exposure to poor air quality has on health as well as the other damage that this exposure causes. Monetary valuation of these impacts can follow, as appropriate. This assessment underpins the choice of air quality improvement targets. Agencies responsible for attaining the targets need to be identified or established, roles and responsibilities need to be allocated to competent authorities, and a broader institutional and governance framework needs to be created to enable implementation of the subsequent steps.

The third step is to identify the key sources of direct emissions of $PM_{2.5}$ and emissions of precursor gases that contribute most to the excessive concentrations of secondary pollutants ($PM_{2.5}$ and ozone) in the previously identified hot spots. This step transcends the traditional policy focus on administrative units and sectors and adopts an *airshed approach* to prioritizing emission sources, abatement measures, and policy instruments.

- *Source-apportionment studies* use laboratory tests of samples of $PM_{2.5}$ particles suspended in the ambient air together with statistical analyses to identify the type of emission source the particles came from, such as vehicles, households, industry, or waste burning, but not necessarily their location. Such studies include sources of direct emissions of $PM_{2.5}$ and, importantly, emissions of their precursors such as SO_2, NO_X, volatile organic compounds, or ammonia.

- For the sources that the source-apportionment studies found critical to air quality, *inventories of emission sources* need to be established, mapping their location, capacity, load profile, and type and amount of fuel use, as well as key source characteristics such as age, stack height, combustion technology, pollution control equipment, and so on.

- *Airshed pollution dispersion models* trace the transport of pollutants from their sources to and within the target airshed; these additional studies capture the formation of secondary pollutants in the airshed (especially $PM_{2.5}$ and ozone). These results then need to be mapped to data on regional population density so that policy interventions can be targeted at those sources that contribute most to population exposure. Many such models are available and have been used for decades (see, for example, Zhou et al. 2006). Such models must consider local geography, topography, atmospheric chemistry, and meteorological conditions to provide useful guidance for achieving agreement on priority abatement measures and the policies to induce them. Model simulations can be additionally validated by field sampling and laboratory tests used for source-apportionment studies.

The fourth step involves assessing the costs and abatement potential of available technical abatement measures that could reduce the population exposure to air pollution in the airshed. The impact of such measures on climate forcing can be estimated at this stage. Synergies and trade-offs between priority measures to mitigate air pollution and measures to mitigate climate change are identified. Measures with high climate co-benefits can be prioritized if they also contribute to health benefits or at least do not deteriorate air quality. The assessment of choices should consider the incremental health effects and premature deaths, incremental costs, and capacity and financing constraints. Equally important are the social and distributional impacts of prioritizing air pollution measures with and without climate co-benefits, including the impact on consumer prices and energy poverty. Before making decisions on capital-intensive air pollution measures, policy makers should order an independent assessment of the economic and social risks of future premature retirement of fossil fuel assets equipped with expensive end-of-pipe pollution-abatement technologies and associated contingent fiscal liabilities.

The fifth step is to design, implement, and enforce the integrated package of coherent policy incentives for firms and households to implement the abatement measures prioritized earlier. Integrated means that the mix of policy instruments needs to encourage economic agents to optimize investment and behavioral decisions, considering both the short-term health impacts of air pollution and the long-term impacts of climate change. Integration also means creatively designing comprehensive mixes of direct regulations (such as emission performance standards, Best Available Technique requirements, or zoning requirements) along with economic and fiscal instruments. Such integration should (1) give firms and households adequate flexibility to achieve air quality objectives at the least cost and (2) encourage innovation and discovery of new, creative abatement measures. If air pollution policies have the potential to induce abatement measures that increase GHG emissions, then additional climate policy efforts need to be identified. Likewise, strengthening air pollution instruments will be necessary if more-ambitious climate policy instruments (to, for example, raise carbon prices) risk increasing air pollution and adverse health impacts. Calibration of the policy mix and the level of ambition of air pollution and climate policy instruments through a dynamic process of review and adjustment is an art rather than a science and must be tailored to the local conditions and political economy dynamics discussed in this report.

It is essential for firms and households to face incentives to reduce both air pollution and GHG emissions. Such coherent incentives help society make the choice between making fossil fuel–based activities cleaner or leapfrogging to new technologies that are free of fossil fuels. These choices will differ by airshed and associated health hazards. Affordability and access to finance are important factors shaping policy choices, especially in low- and middle-income countries.

The World Bank Independent Evaluation Group's report (IEG 2017) called for increased World Bank lending on air pollution because it is responsible for the highest share of deaths caused by any polluting activities. The Independent Evaluation Group also recommended that air pollution interventions be integrated systematically with interventions to mitigate climate change. The present report proposes a practical, science and experience-based approach to operationalizing this recommendation. International development institutions face a challenge to rethink their financing policies to help developing countries proactively manage the synergies and trade-offs between the risks of air pollution and the risks of climate change, rather than chasing co-benefits and always prioritizing win-win measures, especially for domestic heating and cooking in lower-income countries and communities.

NOTES

1. "Clean Air Act 1993" (https://www.legislation.gov.uk/ukpga/1993/11/pdfs/ukpga_19930011 _en.pdf).
2. For the United States, see https://www.epa.gov/mercury/environmental-laws-apply -mercury#CleanAirAct. The concept of a "technique" in EU law is broader than technology. It includes both "the technology used *and* the way in which the installation is designed, built, maintained, operated and decommissioned." https://ec.europa.eu/environment /legal/law/1/module_2_18.htm.
3. The difference between emission taxes and emission charges or fees is important, especially in civil code countries. A tax is defined as compulsory, unrequited payments to general government (OECD 2018). Taxes are unrequited in the sense that benefits provided by governments to taxpayers are not normally in proportion to the payments made by taxpayers. A charge or a fee is a compulsory payment for which a payer expects a reciprocal service from the government, such as a permit to emit a quantity of a specified pollutant, so it is akin to a payment for service. The revenues from charges and fees often go to the agency that issues emission permits or to a dedicated fund, which may not be fully integrated into the state budget. Therefore, the Swedish NO_x charge is not a tax, and in Eastern European countries, environmental charges are legally defined as "payments for the use of the environment."
4. "After Worst Smog in 11 years, Mexico City Braces for More," CBS News, March 18, 2016 (https://www.cbsnews.com/news/after-worst-smog-in-11-years-mexico-city-braces-for -more/).

REFERENCES

Agee, M. D., S. E. Atkinson, T. D. Crocker, and J. W. Williams. 2014. "Non-Separable Pollution Control: Implications for a CO_2 Emissions Cap and Trade System." *Resource and Energy Economics* 36 (1): 64–82. https://doi.org/10.1016/j.reseneeco.2013.11.002.

Air Quality Expert Group. 2018. "Air Pollution from Agriculture." Department for Environment, Food and Rural Affairs; Scottish Government; Welsh Government; and Department of the Environment in Northern Ireland. https://uk-air.defra.gov.uk/assets/documents/reports /aqeg/2800829_Agricultural_emissions_vfinal2.pdf.

Ambec, Stefan, and Jessica Coria. 2013. "Prices vs. Quantities with Multiple Pollutants." *Journal of Environmental Economics and Management* 66: 123–40.

Andaloussi, Mehdi Benatiya. 2018. "Clearing the Air: The Role of Technology Adoption in the Electricity Generation Sector." Working Paper, Toulouse School of Economics.

Åström, Stefan, Katarina Yaramenka, Ingrid Mawdsley, Helena Danielsson, Peringe Grennfelt, Annika Gerner, Tomas Ekvalla, and Erik O. Ahlgren. 2017. "The Impact of Swedish SO_2 Policy Instruments on SO_2 Emissions 1990–2012." *Environmental Science & Policy* 77 (November): 32–39.

Baumol, William J., and Wallace E. Oates. 1971. "The Use of Standards and Prices for Protection of the Environment." *Swedish Journal of Economics* 73 (1): 42–54.

Baumol, William J., and Wallace E. Oates. 1988. *The Theory of Environmental Policy*, 2nd edition. Cambridge, UK, and New York, NY: Cambridge University Press.

Bergquist, Ann-Kristin, Kristina Söderholm, Hanna Kinneryd, Magnus Lindmark, and Patrik Söderholm. 2013. "Command-and-Control Revisited: Environmental Compliance and Technological Change in Swedish Industry 1970–1990." *Ecological Economics* 85 (2013): 6–19. https://doi.org/10.1016/j.ecolecon.2012.10.007.

Bertaud, Alain. 2003. "Clearing the Air in Atlanta: Transit and Smart Growth or Conventional Economics?" *Journal of Urban Economics* 54 (3):379–400. doi:10.1016 /S0094-1190(03)00082-2.

Bonilla, Jorge, Jessica Coria, and Thomas Sterner. 2018. "Technical Synergies and Trade-Offs between Abatement of Global and Local Air Pollution." *Environmental and Resource Economics* 70: 191–221. https://doi.org/10.1007/s10640-017-0117-8.

Bovenberg, A. Lans, and Lawrence H. Goulder. 1996. "Optimal Environmental Taxation in the Presence of Other Taxes: General Equilibrium Analysis." *American Economic Review* 86 (4) (September): 985–1000.

Bovenberg, A. Lans, and Ruud A. de Mooij. 1997. "Environmental Levies and Distortionary Taxation: Reply." *American Economic Review*, 87 (1) (March):252–53.

Bovenberg, A. Lans, and F. van der Ploeg. 1994. "Environmental Policy, Public Finance and the Labour Market in a Second-Best World." *Journal of Public Economics* 55: 349–90.

Brännlund, R., and B. Kriström. 2001. "Too Hot to Handle? Benefits and Costs of Stimulating the Use of Biofuels in the Swedish Heating Sector." *Resource and Energy Economics* 23: 343–58.

Bruce, Nigel, Kristin Aunan, and Eva A. Rehfuess. 2017. "Liquefied Petroleum Gas as a Clean Cooking Fuel for Developing Countries: Implications for Climate, Forests, and Affordability." Materials on Development Financing No. 7, KfW, Frankfurt. https://www.kfw-entwick lungsbank.de/PDF/Download-Center/Materialien/2017_Nr.7_CleanCooking_Lang.pdf.

Burnard, Keith, Julie Jiang, Bo Li, Gilbert Brunet, and Franz Bauer. 2014. "Emissions Reduction through Upgrade of Coal-Fired Power Plants: Learning from Chinese Experience." International Energy Agency, Paris.

Burtraw, Dallas, A. Krupnick, K. Palmer, A. Paul, M. Toman, and C. Bloyd. 2003. "Ancillary Benefits of Reduced Air Pollution in the U.S. from Moderate Greenhouse Gas Mitigation Policies in the Electricity Sector." *Journal of Environmental Economics and Management* 45: 650–73.

Burtraw, Dallas, and Sarah Jo Szambelan. 2009. "U.S. Emissions Trading Markets for SO_2 and NO_x." Discussion Paper RFF DP 09-40. Resources for the Future, Washington, DC.

Cameron, Colin, Shonali Pachauri, Narasimha D. Rao, David McCollum, Joeri Rogelj, and Keywan Riahi. 2016. "Policy Trade-Offs between Climate Mitigation and Clean Cook-Stove Access in South Asia." *Nature Energy* 1: 15010. https://doi.org/10.1038/nenergy.2015.10.

Carlson, A., and D. Burtraw, eds. 2019. "Lessons from the Clean Air Act: Building Durability and Adaptability into US Climate and Energy Policy." Cambridge: Cambridge University Press. doi:10.1017/9781108377195.006.

Croitoru, Lelia, J. C. Chang, and A. Kelly. 2020. "The Cost of Air Pollution in Lagos." World Bank, Washington, DC. https://openknowledge.worldbank.org/handle/10986/33038.

Dworakowska, Anna, Andrzej Guła, Magdalena Kozłowska, Dominika Lalik-Budzewska, Ewa Lutomska, Natalia Matyasik, and Łukasz Pytliński. 2018. "Poza Kontrolą—Analiza Systemu Kontroli Palenisk Domowych." *Krakowski Alarm Smogowy*. https://obserwatoriumdemokracji. pl/wp-content/uploads/2018/11/PozaKontrola_Raport_wersja_elektroniczna.pdf.

EEA (European Environmental Agency). 2019. *Assessing the Effectiveness of EU Policy on Large Combustion Plants in Reducing Air Pollutant Emissions*. EEA Report 07/2019. Luxembourg: Publications Office of the European Union.

Ellerman, Denny A., Paul L. Joskow, and David Harrison Jr. 2003. "Emissions Trading in the U.S.: Experience, Lessons, and Considerations for Greenhouse Gases." Prepared for Pew Center on Global Climate Change, Arlington, VA.

Englert, Dominik, Andrew Losos, Carlo Raucci, and Tristan Smith. 2021. "The Role of LNG in the Transition toward Low- and Zero-Carbon Shipping." World Bank, Washington, DC. https://openknowledge.worldbank.org/handle/10986/35437.

EPRI (Electric Power Research Institute). 2020. "Review of 1.5°C and Other Newer Global Emissions Scenarios." EPRI, Palo Alto, CA. https://www.epri.com/research/products /3002018053.

European Commission. 2021. "Proposal for a COUNCIL DIRECTIVE Restructuring the Union Framework for the Taxation of Energy Products and Electricity (recast)." July 14, 2021. COM(2021)563 final. European Commission, Brussels.

Fuller, Gary. 2018. *The Invisible Killer: The Rising Global Threat of Air Pollution—and How We Can Fight Back*. Brooklyn, NY: Melville House.

Goulder, L. H. 1995. "Environmental Taxation and the 'Double Dividend': A Reader's Guide." *International Tax and Public Finance* 2 (2): 157–83.

Greenstone, Michael, Rohini Pande, Nicholas Ryan, and Anant Sudarshan. 2019. "The Surat Emissions Trading Scheme: A First Look at the World's First Particulate Trading System." Energy Policy Institute, University of Chicago, Chicago, IL. https://epic.uchicago.edu

/research/the-surat-emissions-trading-scheme-a-first-look-at-the-worlds-first-particulate
-trading-system/.

Hamilton, Kirk, Milan Brahmbhatt, Nicholas Bianco, and Jiemei Liu. 2017. "Multiple Benefits
from Climate Mitigation: Assessing the Evidence." Grantham Research Institute on Climate
Change and the Environment, London School of Economics, London. https://www
.researchgate.net/publication/320867824_Multiple_Benefits_from_Climate_Change
_Mitigation_Assessing_the_Evidence.

Hao, Feng. 2020. "New Pricing Could Spell Trouble for China's Coal Sector." China Dialogue.
https://chinadialogue.net/en/energy/11759-new-pricing-could-spell-trouble-for-china-s
-coal-sector/.

Hart, Melanie, Luke Bassett, and Blaine Johnson. 2017a. *Everything You Think You Know about
Coal in China Is Wrong*. Center for American Progress, Washington, DC. https://www
.americanprogress.org/issues/green/reports/2017/05/15/432141/everything-think-know
-coal-china-wrong/.

Hart, Melanie, Luke Bassett, and Blaine Johnson. 2017b. "Research Note on U.S. and Chinese
Coal-Fired Power Data: Assessing Combustion Technology, Efficiency, and Emissions."
Center for American Progress, Washington, DC. https://cdn.americanprogress.org/content
/uploads/2017/05/16113214/ChinaCoal-ResearchNote2.pdf?_ga=2.258286915
.890112332.1599332435-2055023290.1598822363.

Heger, Martin, David Wheeler, Gregor Zens, and Craig Meisner. 2019. "Motor Vehicle Density
and Air Pollution in Greater Cairo: Fuel Subsidy Removal and Metro Line Extension and
Their Effect on Congestion and Pollution." World Bank, Washington, DC. https://doi
.org/10.1596/32512.

IEA (International Energy Agency). 2017. *Energy Access Outlook 2017: From Poverty to Prosperity*.
Paris: IEA. https://iea.blob.core.windows.net/assets/9a67c2fc-b605-4994-8eb5
-29a0ac219499/WEO2017SpecialReport_EnergyAccessOutlook.pdf.

IEA (International Energy Agency). 2020. "India: Only 1 per Cent Coal Power Plant Capacity
Has FGD, 72 Per Cent Yet to Award Bids." International Center for Sustainable Carbon,
London. https://www.iea-coal.org/india-only-1-per-cent-coal-power-plant-capacity-has
-fgd-72-per-cent-yet-to-award-bids/.

IMF (International Monetary Fund). 2019. "Fiscal Policies for Paris Climate Strategies—
from Principle to Practice." Fiscal Affairs Dept. Publication, Policy Paper 19/010, IMF,
Washington, DC.

Independent Evaluation Group. 2017. "Toward a Clean World for All: An IEG Evaluation of the
World Bank Group's Support to Pollution Management." World Bank, Washington, DC.
https://openknowledge.worldbank.org/handle/10986/28957.

International Transport Forum. 2017. *Strategies for Mitigating Air Pollution in Mexico City:
International Best Practice*. Paris: International Transport Forum. https://www.itf-oecd
.org/sites/default/files/docs/air-pollution-mitigation-strategy-mexico-city.pdf.

Iyu, Wanning, Yuan Li, Dabo Guan, Hongyan Zhao, Qiang Zhang, and Zhu Liu. 2016. "Driving
Forces of Chinese Primary Air Pollution Emissions: An Index Decomposition Analysis."
Journal of Cleaner Production 133 (1): 136–44. doi:10.1016/j.jclepro.2016.04.093.

Kheiravar, Khaled H. 2019. "Economic and Econometric Analyses of the World Petroleum
Industry, Energy Subsidies, and Air Pollution." PhD thesis, UC Davis Institute of
Transportation Studies. https://escholarship.org/uc/item/3gj151w9.

Klausbruckner, C., H. Annegarn, L. R. F. Henneman, and P. Rafaj. 2016. "A Policy Review of
Synergies and Trade-Offs in South African Climate Change Mitigation and Air Pollution
Control Strategies." *Environmental Science and Policy* 57: 70–78. https://doi.org/10.1016
/j.envsci.2015.12.001.

Kypridemos, Chris, Elisa Puzzolo, Borgar Aamaas, Lirije Hyseni, Matthew Shupler, Kristin
Aunan, and Daniel Pope. 2020. "Health and Climate Impacts of Scaling Adoption of
Liquefied Petroleum Gas (LPG) for Clean Household Cooking in Cameroon: A Modeling
Study." *Environmental Health Perspectives* 128 (4): 47001. doi:10.1289/EHP4899.

Lee, Carrie M., Chelsea Chandler, Michael Lazarus, and Francis X. Johnson. 2013. "Assessing
the Climate Impacts of Cookstove Projects: Issues in Emissions Accounting." Working
Paper 2013-01, Stockholm Environment Institute.

Li, Tao, Yimiao Song, and Jing Shen. 2019. "Clean Power Dispatching of Coal-Fired Power Generation in China Based on the Production Cleanliness Evaluation Method." *Sustainability* 11: 6778. doi:10.3390/su11236778.

Liu, Antung. 2013. "Tax Evasion and Optimal Environmental Taxes." *Journal of Environmental Economics and Management* 66 (3): 656–70. https://doi.org/10.1016/j.jeem.2013.06.004.

Lvovsky, Kseniya, Gordon Hughes, David Maddison, Bart Ostro, and David Pearce. 2000. "Environmental Costs of Fossil Fuels: A Rapid Assessment Method with Application to Six Cities." Environment Department Paper 78, Pollution Management Series, World Bank, Washington, DC. https://openknowledge.worldbank.org/handle/10986/18303.

Massetti, Emanuele, Marilyn A. Brown, Melissa Lapsa, Isha Sharma, James Bradbury, Colin Cunliff, and Yufei Li. 2017. *Environmental Quality and the U.S. Power Sector: Air Quality, Water Quality, Land Use and Environmental Justice*. Oak Ridge, TN: Oak Ridge National Laboratory. https://info. ornl.gov/sites/ publications/ files/Pub60561.pdf

McNevin, Thomas F. 2016. "Recent Increases in Nitrogen Oxide (NO_x) Emissions from Coal-Fired Electric Generating Units Equipped with Selective Catalytic Reduction." *Journal of the Air & Waste Management Association* 66 (1): 66–75. doi:10.1080/10962247.2015.1112317.

Mooij, de R. A. 1999. "The Double Dividend of an Environmental Tax Reform." In *Handbook of Environmental and Resource Economics*, edited by Jeroen C. J. M. van den Bergh, chapter 20. Cheltenham, UK: Edward Elgar Publishing. https://doi.org/10.4337/9781843768586.

Myllyvirta, Lauri, Shuwei Zhang, and Xinyi Shen. 2020. "Will China Build Hundreds of New Coal Plants in the 2020s?" Cabon Brief. https://www.carbonbrief.org/analysis-will-china -build-hundreds-of-new-coal-plants-in-the-2020s.

NASA (National Aeronautics and Space Administration). 2017. "China's Sulfur Dioxide Emissions Drop, India's Grow Over Last Decade." NASA website. https://www.nasa.gov /feature/goddard/2017/chinas-sulfur-dioxide-emissions-drop-indias-grow-over-last -decade.

Newell, Richard G., and Kristian Rogers. 2003. "The U.S. Experience with the Phasedown of Lead in Gasoline." Discussion Paper, Resources for the Future, Washington, DC. https:// web.mit.edu/ckolstad/www/Newell.pdf.

OECD (Organisation for Economic Co-operation and Development). 2009. *Ensuring Environmental Compliance: Trends and Good Practices*. Paris: OECD.

OECD (Organisation for Economic Co-operation and Development). 2013. "The Swedish Tax on Nitrogen Oxide Emissions: Lessons in Environmental Policy Reform." Environment Policy Paper 2, OECD, Paris.

OECD. 2018. *Revenue Statistics 2018*. Paris: OECD Publishing. https://doi.org/10.1787/rev _stats-2018-en.

OECD (Organisation for Economic Co-operation and Development). 2019. *Taxing Energy Use 2019: Using Taxes for Climate Action*. Paris: OECD Publishing. https://doi.org/10.1787 /058ca239-en.

OECD (Organisation for Economic Co-operation and Development). 2021. Policy Instruments for the Environment (PINE) database. OECD, Paris. https://www.oecd.org/env/indicators -modelling-outlooks/policy-instrument-database/.

Pachauri, Shonali, Narasimha D. Rao, and Colin Cameron. 2018. "Outlook for Modern Cooking Energy Access in Central America." *PLoS One* 13 (6): e0197974. doi:10.1371/journal .pone.0197974.

Peszko, Grzegorz. 1999. "Polish Experience with Environmental Fees, Fines and Taxes, and Simulations of Some Economic Effects of Elements of Green Tax Reform Using the Computable General Equilibrium Model." In *Green Budget Reform in Europe: Countries at the Forefront*, edited by Kai Schlegelmilch, 127–48. Berlin: Springer.

Peszko, Grzegorz, Simon John Black, Alexandrina Platonova-Oquab, Dirk Heine, and Govinda R. Timilsina. 2019. *Environmental Fiscal Reform in Morocco: Options and Pathways*. Washington, DC: World Bank Group. http://documents.worldbank.org/curated/en/450501560190965482 /Environmental-Fiscal-Reform-in-Morocco-Options-and-Pathways.

Peszko, Grzegorz, Dominique van der Mensbrugghe, Alexander Golub, John Ward, Dimitri Zenghelis, Cor Marijs, Anne Schopp, John A. Rogers, and Amelia Midgley. 2020.

Diversification and Cooperation in a Decarbonizing World: Climate Strategies for Fossil Fuel–Dependent Countries. Washington, DC: World Bank. https://openknowledge.worldbank.org/handle/10986/34011.

Petetin, Hervé, Jean Sciare, Michael Bressi, Valérie Gros, Amandine Rosso, Olivier Sanchez, Roland Sarda-Estève, Jean-Eudes Petit, and Matthias Beekmann. 2016. "Assessing the Ammonium Nitrate Formation Regime in the Paris Megacity and Its Representation in the CHIMERE Model." *Atmospheric Chemistry and Physics* 16: 10419–40. www.atmos-chem-phys.net/16/10419/2016/.

Pigato, Miria A., ed. 2019. *Fiscal Policies for Development and Climate Action.* Washington, DC: World Bank Group. http://documents.worldbank.org/curated/en/340601545406276579/Fiscal-Policies-for-Development-and-Climate-Action.

Pigou, A. C. 1920. *The Economics of Welfare.* London: Macmillan.

Pittel, K., and D. Rübbelke. 2008. "Climate Policy and Ancillary Benefits: A Survey and Integration into the Modelling of International Negotiations on Climate Change." *Ecological Economics* 68 (1–2): 210–20.

Portugal-Pereira, Alexandre Koberle, André F. P. Lucena, Pedro R. R. Rochedo, Mariana Império, Ana Monteiro Carsalade, Roberto Schaeffer, and Peter Rafaj. 2018. "Interactions between Global Climate Change Strategies and Local Air Pollution: Lessons Learnt from the Expansion of the Power Sector in Brazil." *Climatic Change* 148: 293–309. https://doi.org/10.1007/s10584-018-2193-3.

Rafaj, Peter, and Markus Amann. 2018. "Decomposing Air Pollutant Emissions in Asia: Determinants and Projections." *Energies* 11: 1299. doi:10.3390/en11051299.

Rafaj, Peter, Markus Amann, and José G. Siri. 2014. "Factorization of Air Pollutant Emissions: Projections versus Observed Trends in Europe." *Science of the Total Environment* 494–495: 272–82. doi:10.1016/j.scitotenv.2014.07.013.

Rafaj, Peter, M. Amann, J. Siri, and H. Wuester. 2014. "Changes in European Greenhouse Gas and Air Pollutant Emissions 1960–2010: Decomposition of Determining Factors." *Climatic Change* 124: 477–504.

Regens, James L., and Robert W. Rycroft. 1988. *The Acid Rain Controversy.* Pittsburgh, PA: University of Pittsburgh Press.

Republic of Serbia. 2019. "Draft Low Carbon Development Strategy with Action Plan." GFA Consulting Group GmbH, Hamburg. https://balkangreenenergynews.com/wp-content/uploads/2020/01/Low-Carbon-Development-Strategy-with-Action-plan-Serbia_eng.pdf.

Rogelj, J., D. L. McCollum, and K. Riahi. 2013. "The UN's 'Sustainable Energy for All Initiative' Is Compatible with a Warming Limit of 2°C." *Nature Climate Change* 3: 545–51.

Ryani, Lisa, Ivan Petrovi, Andrew Kellyii, Yulu Guoi, and Sarah La Monaca. 2019. "An Assessment of the Social Costs and Benefits of Vehicle Tax Reform in Ireland." Environment Working Papers No. 153, OECD, Paris.

Schmalensee, Richard, and Robert N. Stavins. 2012. "The SO_2 Allowance Trading System: The Ironic History of a Grand Policy Experiment." *Journal of Economic Perspectives* 27 (1): 103–22.

SEPA (Swedish Environmental Protection Agency) and the Swedish Energy Agency. 2007. "Economic Instruments in Environmental Policy." Report 5678. Stockholm.

Singh, Mandvi. 2017. "How China Is Cleaning the Highly-Polluting Coal Power Sector." Down to Earth, accessed August 8, 2020, https://www.downtoearth.org.in/news/air/china-cleans-up-its-act-58505.

Slovic, A. D., M. A. de Oliveira, J. Biehl, and H. Ribeiro. 2016. "How Can Urban Policies Improve Air Quality and Help Mitigate Global Climate Change: A Systematic Mapping Review." *Journal of Urban Health* 93: 73–95. https://doi.org/10.1007/s11524-015-0007-8.

Statistics Poland. 2020. *Economic Aspects of Environmental Protection.* Warsaw: Statistics Poland.

Stern, Nicholas. 2008. "The Economics of Climate Change." *American Economic Review* 98 (2): 1–37. http://www.jstor.org/stable/29729990.

Sterner, Thomas. 2007. "Fuel Taxes: An Important Instrument for Climate Policy." *Energy Policy* 35 (6): 3194–202. doi:10.1016/j.enpol.2006.10.025.

Sterner, Thomas, and Lena Hoglund Isaksson. 2006. "Refunded Emission Payments Theory, Distribution of Costs, and Swedish Experience of NO$_x$ Abatement." *Ecological Economics* 57: 93–106.

Strasert, Brian, Su Chen Teh, and Daniel S. Cohan. 2019. "Air Quality and Health Benefits from Potential Coal Power Plant Closures in Texas." *Journal of the Air & Waste Management Association* 69: (3) 333–50. doi:10.1080/10962247.2018.1537984.

Svenningsen, Lea Skræp, Liv Lærke Hansen, Michael Munk Sørensen, Emelie von Bahr, Hrafnhildur Bragadóttir, Kennet Christian Uggeldahl, Hanne Søiland, et al. 2019. *The Use of Economic Instruments in Nordic Environmental Policy 2014–2017*. Copenhagen: Nordic Council of Ministers.

Swedish Forest Agency. 2018. "Market Statement 2018—Sweden." UNECE Timber Committee Market Discussion, Vancouver, Canada, November 5–9.

Tan-Soo, Jie-Sheng, Zhang Xiao-Bing, Qin Ping, and Xie Lunyu. 2019. "Using Electricity Prices to Curb Industrial Pollution." *Journal of Environmental Management* 248: 109252.

Thaler, Richard H., and Cass R. Sunstein. 2008. *Nudge: Improving Decisions about Health, Wealth, and Happiness*. New Haven, CT: Yale University Press.

Thompson, Gavin. 2020. "Is China Embarking on a Major Expansion of Coal-Fired Power Generation?" Wood Mackenzie. https://www.woodmac.com/news/opinion/is-china -embarking-on-a-major-expansion-of-coal-fired-power-generation/.

US EPA (US Environmental Protection Agency). 2009. "The NO$_x$ Budget Trading Program: 2008 Emission, Compliance, and Market Analyses." US EPA, Washington, DC. https://www .epa.gov/sites/default/files/2015-08/documents/nbp_2008_ecm_analyses.pdf.

US EPA (US Environmental Protection Agency). 2010. "Clean Air Interstate Rule: 2009 Emission, Compliance and Market Analyses." US EPA, Washington, DC. https://www.epa .gov/sites/production/files/2015-08/documents/cair09_ecm_analyses.pdf.

Weber, Paige. 2021. "Dynamic Responses to Carbon Pricing in the Electricity Sector." Unpublished, University of North Carolina, Chapel Hill, NC. https://www.paige-weber .com/uploads/1/2/6/7/126784653/weber_ghg_wpjan2021.pdf.

Wei, Wenjing, Pelle Mellin, Weihong Yang, Chuan Wang, Anders Hultgren, and Hassan Salman. 2013. "Utilization of Biomass for Blast Furnace in Sweden: Report I: Biomass Availability and Upgrading Technologies." KTH Industrial Engineeering and Management. doi:10.13140/RG.2.2.13641.39522.

Weitzman, M. L. 1974. "Prices vs. Quantities." *Review of Economic Studies* 41 (4): 477–91.

World Bank. 2009. "Mexico—Sustainable Transport and Air Quality Project." World Bank, Washington, DC. http://documents.worldbank.org/curated/en/452081468049873856 /Mexico-Sustainable-Transport-and-Air-Quality-Project.

World Bank. 2020a. "Hebei Air Pollution Prevention and Control Program: Implementation Completion and Results Report." World Bank, Washington, DC. http://documents1.world bank.org/curated/en/746591593402580377/pdf/China-Hebei-Air-Pollution-Prevention -and-Control-Project.pdf.

World Bank. 2020b. *State and Trends of Carbon Pricing 2020*. Washington, DC: World Bank. https://openknowledge.worldbank.org/handle/10986/33809.

World Bank. 2021. "Are All Air Pollution Particles Equal? How Constituents and Sources of Fine Air Pollution Particles (PM$_{2.5}$) Affect Health." World Bank, Washington, DC.

Yan, S., and G. Wu. 2017. "SO$_2$ Emissions in China: Their Network and Hierarchical Structures." *Scientific Reports* 7: 46216. https://doi.org/10.1038/srep46216.

Zhao, Xiuling, Weiqi Zhou, Lijian Han, and Dexter Locke. 2019. "Spatiotemporal Variation in PM$_{2.5}$ Concentrations and Their Relationship with Socioeconomic Factors in China's Major Cities." *Environment International* 133 Part A (December): 105145.

Zhou, Ying, Jonathan I. Levy, John S. Evans, and James K. Hammitt. 2006. "The Influence of Geographic Location on Population Exposure to Emissions from Power Plants throughout China." *Environment International* 32 (3): 365–73. https://chinaproject.harvard.edu /publications/influence-geographic-location-population-exposure-emissions-power -plants-throughout-0.